PERIODIC
TABLE

Memorize the elements of the periodic table with its symbols and spelling

Yash'al Ahmed Abdul Sattar
Aminath Sharahath

To order additional copies of this book, contact
Toll Free +65 3165 7531 (Singapore)
Toll Free +60 3 3099 4412 (Malaysia)
www.partridgepublishing.com/singapore
orders.singapore@partridgepublishing.com

ISBN
978-1-5437-6975-3 (sc)
978-1-5437-6976-0 (hc)
978-1-5437-6977-7 (e)

Library of Congress Control Number: 2022907732

Print information available on the last page.

04/13/2022

PARTRIDGE

HOW TO USE THIS BOOK

Having the periodic table memorized along with chemical symbols and their spelling is going to be very helpful in your science class.

This book will help you memorize the periodic table, the symbols, and the spelling in a fun way. You won't even know you are learning. By the time you are done with the set of puzzles, you'll be amazed at how much you know within such a short period of time, while you are having fun!

The puzzles are very simple, and easy. It is divided into 12 groups according to the order in the periodic table. Each group will have 10 elements except the last group, which has 8 elements. There are 5 crossword puzzles in each group to help you with your learning process.

All you need to do is use the chemical symbols of the elements to fill in the crossword puzzle with the element name. The first 3 puzzles in each set will give you the word bank. They are color coded to help you. By the end of the first 3 puzzles, you will be familiar with the spelling so the word bank of the last 2 puzzles in the set will be replaced with a very simple fill in the blanks. They are color coded, so it is still very easy, but challenging enough to test your memory.

EXAMPLE 1 (filling the clues under 'DOWN')

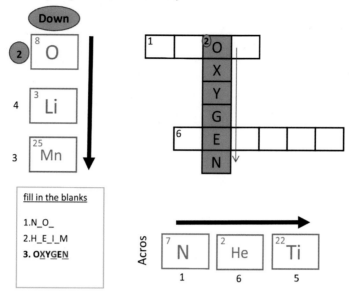

Down

2 | 8 O

4 | 3 Li

3 | 25 Mn

fill in the blanks

1.N_O_
2.H_E_I_M
3. OXYGEN

Across

7 N — 1
2 He — 6
22 Ti — 5

EXAMPLE 2 (filling the clues under 'ACROSS')

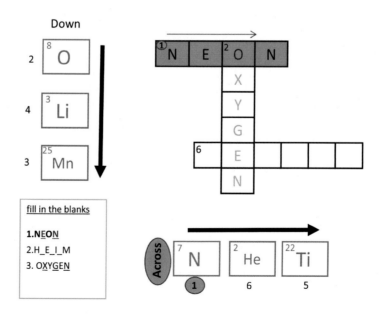

Down

2 | 8 O

4 | 3 Li

3 | 25 Mn

fill in the blanks

1.NEON
2.H_E_I_M
3. OXYGEN

Across

7 N — 1
2 He — 6
22 Ti — 5

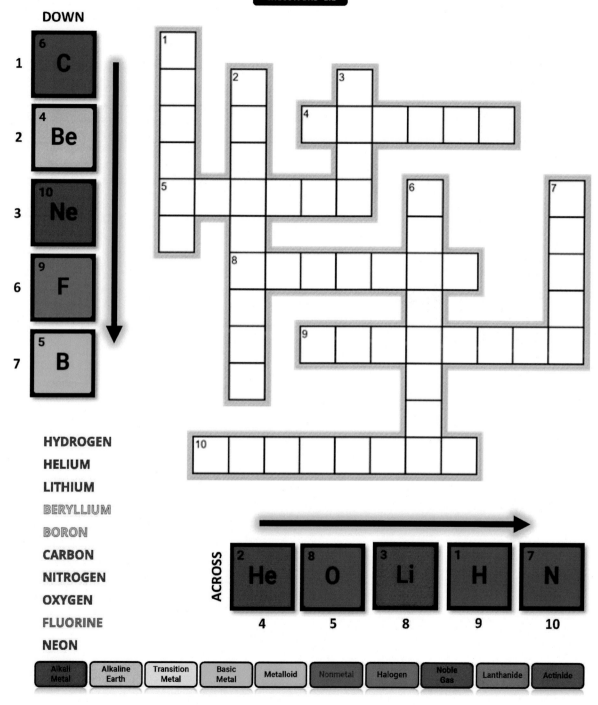

CROSSWORD 1.1

DOWN

1. C (6)
2. Be (4)
3. Ne (10)
6. F (9)
7. B (5)

HYDROGEN
HELIUM
LITHIUM
BERYLLIUM
BORON
CARBON
NITROGEN
OXYGEN
FLUORINE
NEON

ACROSS

4. He (2)
5. O (8)
8. Li (3)
9. H (1)
10. N (7)

Alkali Metal | Alkaline Earth | Transition Metal | Basic Metal | Metalloid | Nonmetal | Halogen | Noble Gas | Lanthanide | Actinide

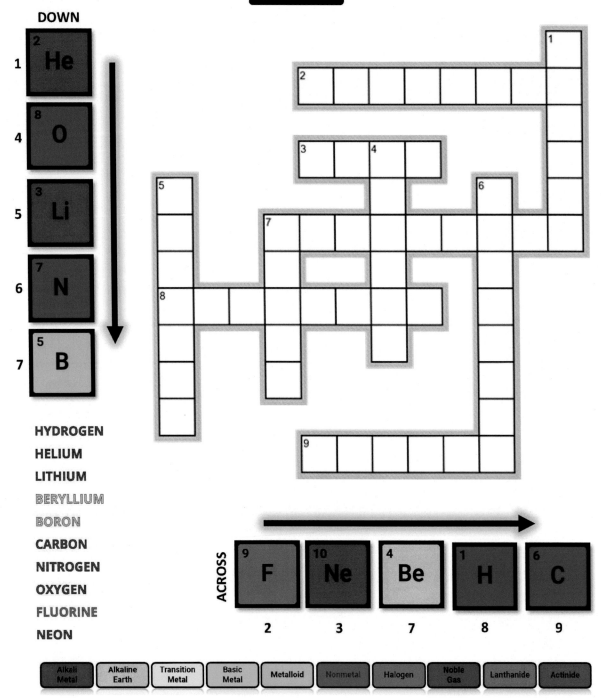

DOWN

1 He 2

4 O 8

5 Li 3

6 N 7

7 B 5

HYDROGEN
HELIUM
LITHIUM
BERYLLIUM
BORON
CARBON
NITROGEN
OXYGEN
FLUORINE
NEON

ACROSS

F 9 | Ne 10 | Be 4 | H 1 | C 6

2 | 3 | 7 | 8 | 9

Alkali Metal | Alkaline Earth | Transition Metal | Basic Metal | Metalloid | Nonmetal | Halogen | Noble Gas | Lanthanide | Actinide

CROSSWORD 1.3

DOWN

#		#
2	3 **Li**	
3	6 **C**	
4	1 **H**	
5	5 **B**	
6	9 **F**	

HYDROGEN
HELIUM
LITHIUM
BERYLLIUM
BORON
CARBON
NITROGEN
OXYGEN
FLUORINE
NEON

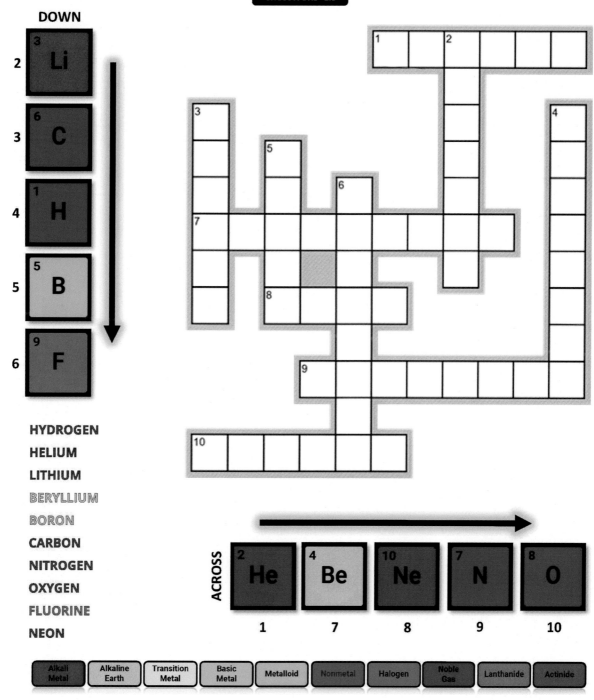

ACROSS

2 **He**	4 **Be**	10 **Ne**	7 **N**	8 **O**
1	7	8	9	10

Alkali Metal	Alkaline Earth	Transition Metal	Basic Metal	Metalloid	Nonmetal	Halogen	Noble Gas	Lanthanide	Actinide

DOWN

1 He (2)
4 O (8)
5 N (7)
7 Li (3)
9 B (5)

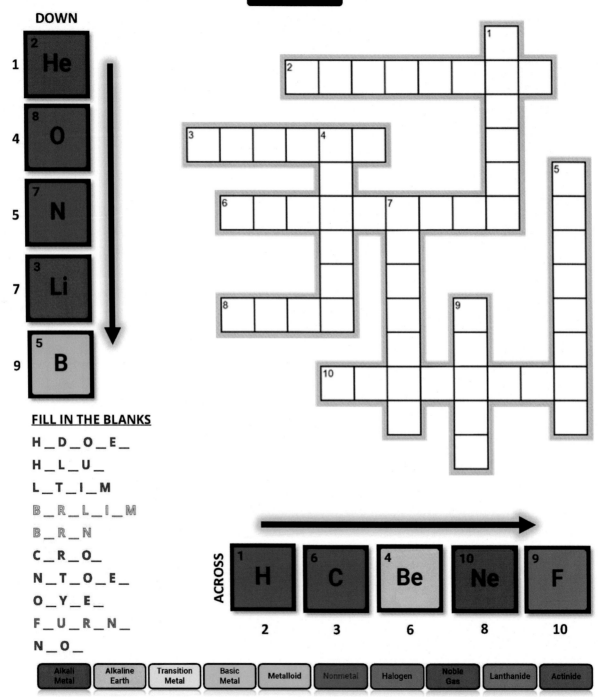

FILL IN THE BLANKS

H _ D _ O _ E _

H _ L _ U _

L _ T _ I _ M

B _ R _ L _ I _ M

B _ R _ N

C _ R _ O _

N _ T _ O _ E _

O _ Y _ E _

F _ U _ R _ N _

N _ O _

ACROSS

1 H (2)
6 C (3)
4 Be (6)
10 Ne (8)
9 F (10)

Alkali Metal | Alkaline Earth | Transition Metal | Basic Metal | Metalloid | Nonmetal | Halogen | Noble Gas | Lanthanide | Actinide

DOWN

2. 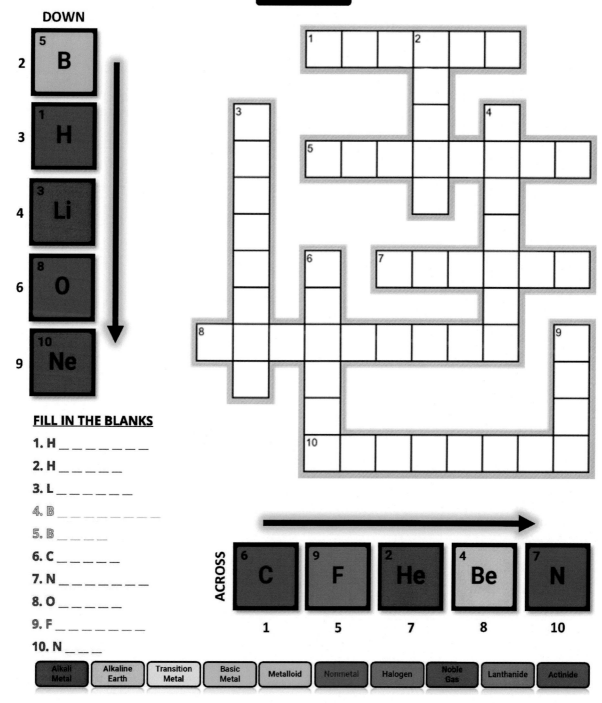 B (5)
3. H (1)
4. Li (3)
6. O (8)
9. Ne (10)

FILL IN THE BLANKS

1. H _ _ _ _ _ _ _ _
2. H _ _ _ _ _ _
3. L _ _ _ _ _ _ _
4. B _ _ _ _ _ _ _ _ _
5. B _ _ _ _
6. C _ _ _ _ _ _
7. N _ _ _ _ _ _ _ _
8. O _ _ _ _ _ _
9. F _ _ _ _ _ _ _ _ _
10. N _ _ _ _

ACROSS

- C (6) — 1
- F (9) — 5
- He (2) — 7
- Be (4) — 8
- N (7) — 10

Alkali Metal | Alkaline Earth | Transition Metal | Basic Metal | Metalloid | Nonmetal | Halogen | Noble Gas | Lanthanide | Actinide

DOWN

1. **K** 19
2. **Na** 11
3. **Al** 13
4. **Ca** 20
7. **Ar** 18

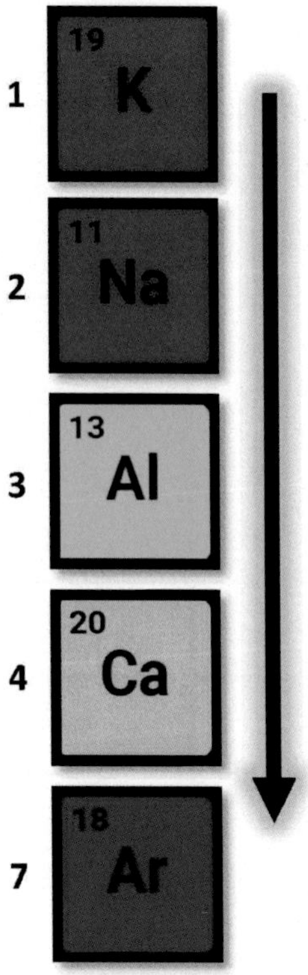

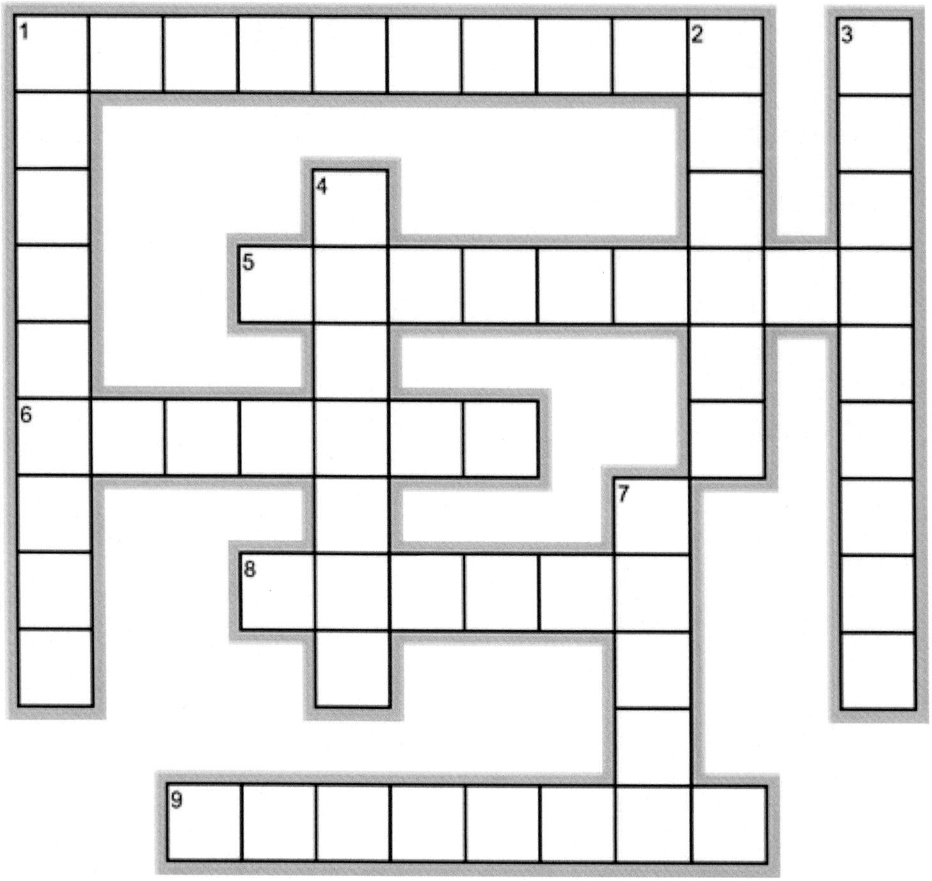

SODIUM
MAGNESIUM
ALUMINIUM
SILICON
PHOSPHORUS
SULFUR
CHLORINE
ARGON
POTASSIUM
CALCIUM

ACROSS

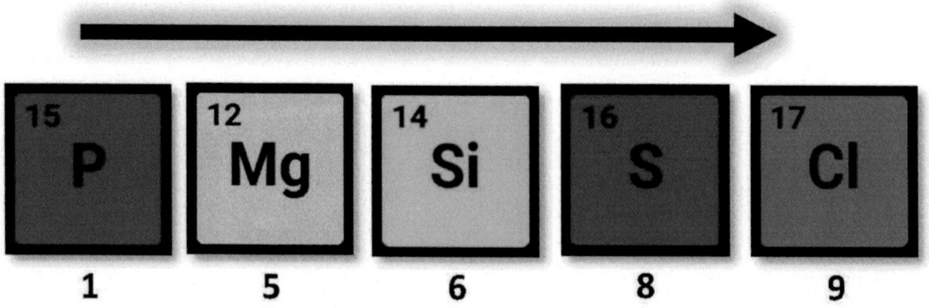

1. **P** 15
5. **Mg** 12
6. **Si** 14
8. **S** 16
9. **Cl** 17

| Alkali Metal | Alkaline Earth | Transition Metal | Basic Metal | Metalloid | Nonmetal | Halogen | Noble Gas | Lanthanide | Actinide |

DOWN

1. 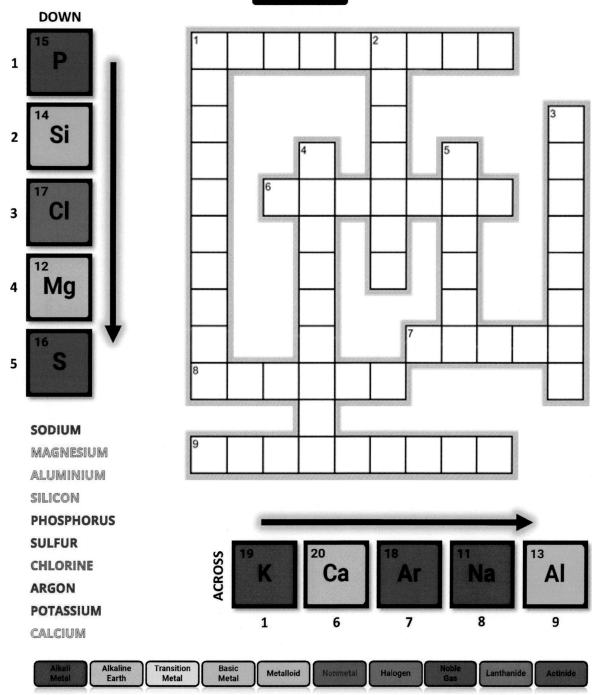 ¹⁵ P
2. ¹⁴ Si
3. ¹⁷ Cl
4. ¹² Mg
5. ¹⁶ S

SODIUM
MAGNESIUM
ALUMINIUM
SILICON
PHOSPHORUS
SULFUR
CHLORINE
ARGON
POTASSIUM
CALCIUM

ACROSS

¹⁹ K	²⁰ Ca	¹⁸ Ar	¹¹ Na	¹³ Al
1	6	7	8	9

Alkali Metal | Alkaline Earth | Transition Metal | Basic Metal | Metalloid | Nonmetal | Halogen | Noble Gas | Lanthanide | Actinide

DOWN

1. ^{14} **Si**
2. ^{15} **P**
4. ^{13} **Al**
5. ^{17} **Cl**
6. ^{20} **Ca**

SODIUM
MAGNESIUM
ALUMINIUM
SILICON
PHOSPHORUS
SULFUR
CHLORINE
ARGON
POTASSIUM
CALCIUM

ACROSS

^{19} K	^{11} Na	^{18} Ar	^{16} S	^{12} Mg
2	3	7	8	9

Alkali Metal · Alkaline Earth · Transition Metal · Basic Metal · Metalloid · Nonmetal · Halogen · Noble Gas · Lanthanide · Actinide

DOWN

1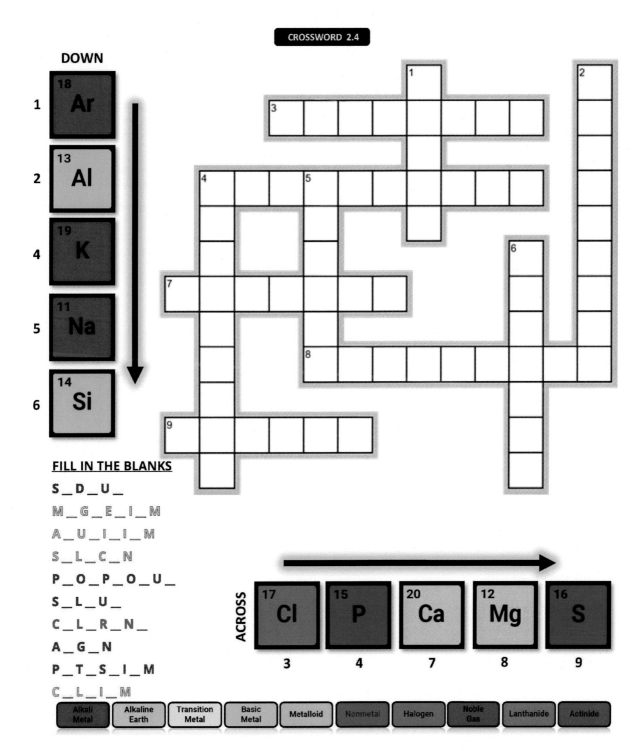
18 Ar

2
13 Al

4
19 K

5
11 Na

6
14 Si

FILL IN THE BLANKS

S _ D _ U _

M _ G _ E _ I _ M

A _ U _ I _ I _ M

S _ L _ C _ N

P _ O _ P _ O _ U _

S _ L _ U _

C _ L _ R _ N _

A _ G _ N

P _ T _ S _ I _ M

C _ L _ I _ M

ACROSS

17 Cl — 3
15 P — 4
20 Ca — 7
12 Mg — 8
16 S — 9

Alkali Metal | Alkaline Earth | Transition Metal | Basic Metal | Metalloid | Nonmetal | Halogen | Noble Gas | Lanthanide | Actinide

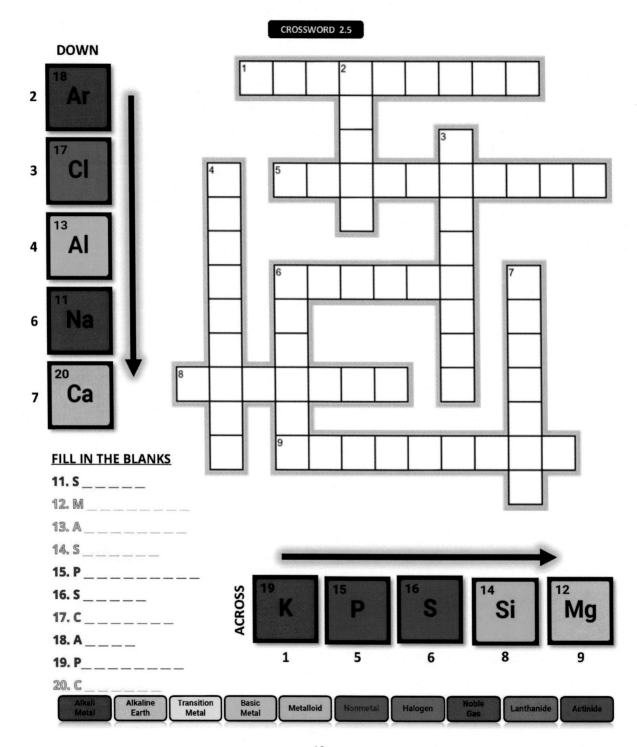

CROSSWORD 2.5

DOWN

2. **Ar** 18
3. **Cl** 17
4. **Al** 13
6. **Na** 11
7. **Ca** 20

FILL IN THE BLANKS

11. S _ _ _ _ _ _
12. M _ _ _ _ _ _ _ _
13. A _ _ _ _ _ _ _ _
14. S _ _ _ _ _ _
15. P _ _ _ _ _ _ _ _ _ _
16. S _ _ _ _ _
17. C _ _ _ _ _ _ _ _
18. A _ _ _ _ _
19. P _ _ _ _ _ _ _ _ _
20. C _ _ _ _ _ _ _

ACROSS

1. **K** 19
5. **P** 15
6. **S** 16
8. **Si** 14
9. **Mg** 12

Alkali Metal | Alkaline Earth | Transition Metal | Basic Metal | Metalloid | Nonmetal | Halogen | Noble Gas | Lanthanide | Actinide

CROSSWORD 3.1

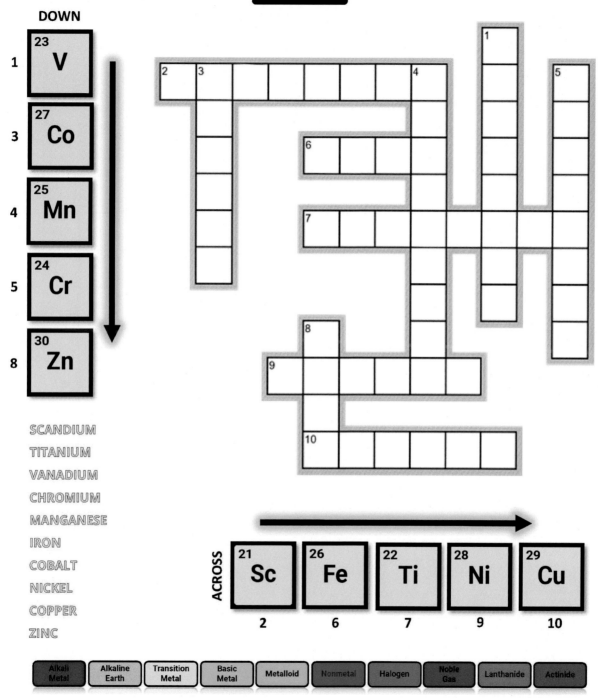

DOWN

1. 23 V
3. 27 Co
4. 25 Mn
5. 24 Cr
8. 30 Zn

SCANDIUM
TITANIUM
VANADIUM
CHROMIUM
MANGANESE
IRON
COBALT
NICKEL
COPPER
ZINC

ACROSS

21 Sc — 2
26 Fe — 6
22 Ti — 7
28 Ni — 9
29 Cu — 10

| Alkali Metal | Alkaline Earth | Transition Metal | Basic Metal | Metalloid | Nonmetal | Halogen | Noble Gas | Lanthanide | Actinide |

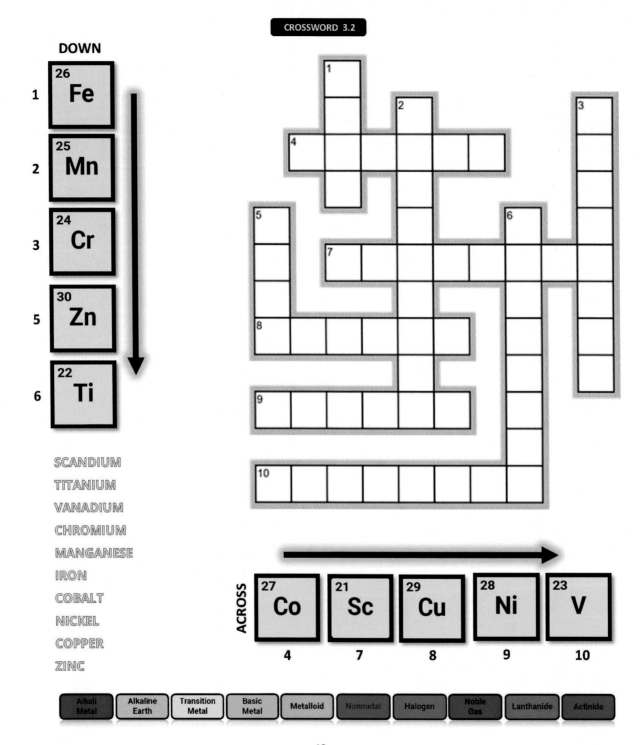

CROSSWORD 3.2

DOWN

1 ²⁶ Fe

2 ²⁵ Mn

3 ²⁴ Cr

5 ³⁰ Zn

6 ²² Ti

SCANDIUM
TITANIUM
VANADIUM
CHROMIUM
MANGANESE
IRON
COBALT
NICKEL
COPPER
ZINC

ACROSS

²⁷ Co   ²¹ Sc   ²⁹ Cu   ²⁸ Ni   ²³ V
4   7   8   9   10

| Alkali Metal | Alkaline Earth | Transition Metal | Basic Metal | Metalloid | Nonmetal | Halogen | Noble Gas | Lanthanide | Actinide |

DOWN

#	Element
2	28 Ni
4	22 Ti
6	23 V
7	30 Zn
8	26 Fe

SCANDIUM
TITANIUM
VANADIUM
CHROMIUM
MANGANESE
IRON
COBALT
NICKEL
COPPER
ZINC

ACROSS

25 Mn	27 Co	29 Cu	21 Sc	24 Cr
1	3	5	9	10

| Alkali Metal | Alkaline Earth | Transition Metal | Basic Metal | Metalloid | Nonmetal | Halogen | Noble Gas | Lanthanide | Actinide |

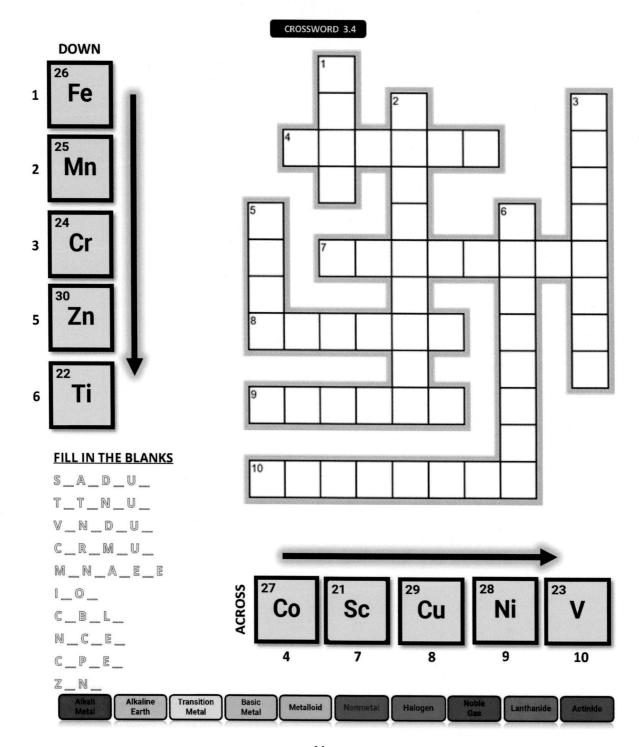

CROSSWORD 3.4

DOWN

1 ²⁶ Fe

2 ²⁵ Mn

3 ²⁴ Cr

5 ³⁰ Zn

6 ²² Ti

FILL IN THE BLANKS

S _ A _ D _ U _
T _ T _ N _ U _
V _ N _ D _ U _
C _ R _ M _ U _
M _ N _ A _ E _ E
I _ O _
C _ B _ L _
N _ C _ E _
C _ P _ E _
Z _ N _

ACROSS

²⁷ Co — 4

²¹ Sc — 7

²⁹ Cu — 8

²⁸ Ni — 9

²³ V — 10

Alkali Metal | Alkaline Earth | Transition Metal | Basic Metal | Metalloid | Nonmetal | Halogen | Noble Gas | Lanthanide | Actinide

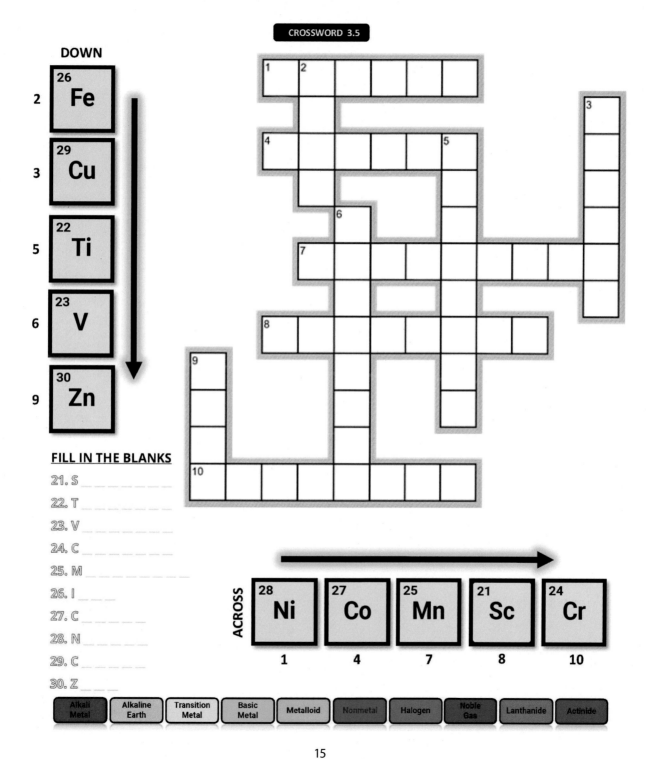

CROSSWORD 3.5

DOWN

2 **Fe** 26

3 **Cu** 29

5 **Ti** 22

6 **V** 23

9 **Zn** 30

FILL IN THE BLANKS

21. S _____

22. T _____

23. V _____

24. C _____

25. M _____

26. I _____

27. C _____

28. N _____

29. C _____

30. Z _____

ACROSS

Ni 28 — 1

Co 27 — 4

Mn 25 — 7

Sc 21 — 8

Cr 24 — 10

Alkali Metal | Alkaline Earth | Transition Metal | Basic Metal | Metalloid | Nonmetal | Halogen | Noble Gas | Lanthanide | Actinide

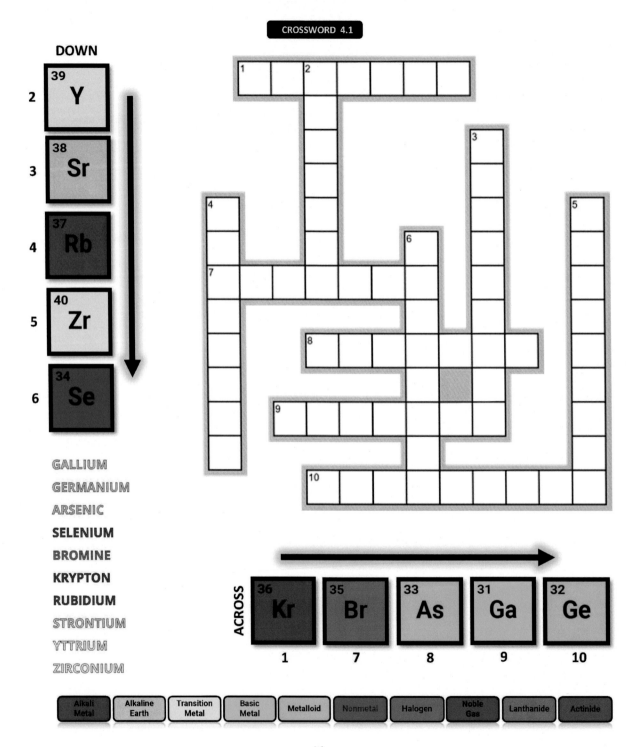

CROSSWORD 4.1

DOWN

2. **Y** (39)
3. **Sr** (38)
4. **Rb** (37)
5. **Zr** (40)
6. **Se** (34)

GALLIUM
GERMANIUM
ARSENIC
SELENIUM
BROMINE
KRYPTON
RUBIDIUM
STRONTIUM
YTTRIUM
ZIRCONIUM

ACROSS

1. **Kr** (36)
7. **Br** (35)
8. **As** (33)
9. **Ga** (31)
10. **Ge** (32)

| Alkali Metal | Alkaline Earth | Transition Metal | Basic Metal | Metalloid | Nonmetal | Halogen | Noble Gas | Lanthanide | Actinide |

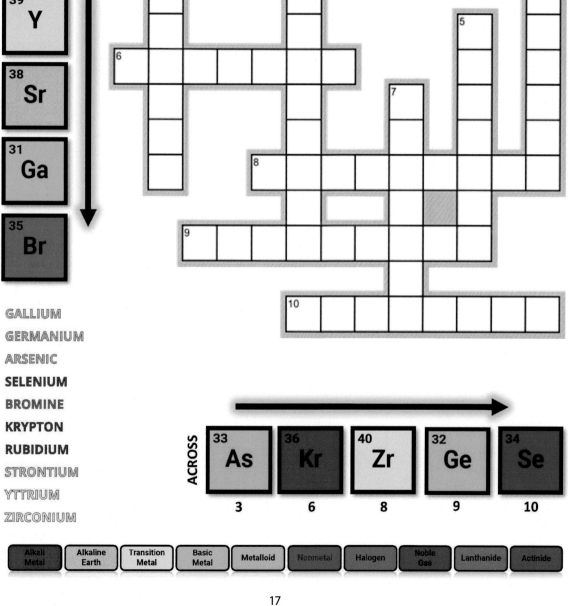

CROSSWORD 4.2

DOWN

#	Symbol
1	37 Rb
2	39 Y
4	38 Sr
5	31 Ga
7	35 Br

GALLIUM
GERMANIUM
ARSENIC
SELENIUM
BROMINE
KRYPTON
RUBIDIUM
STRONTIUM
YTTRIUM
ZIRCONIUM

ACROSS

33 As	36 Kr	40 Zr	32 Ge	34 Se
3	6	8	9	10

| Alkali Metal | Alkaline Earth | Transition Metal | Basic Metal | Metalloid | Nonmetal | Halogen | Noble Gas | Lanthanide | Actinide |

17

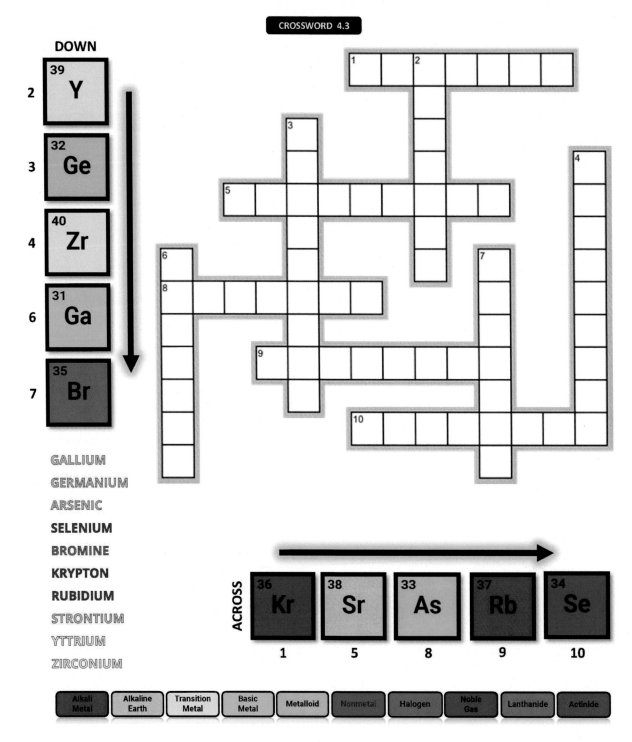

CROSSWORD 4.3

DOWN

2 | 39 Y
3 | 32 Ge
4 | 40 Zr
6 | 31 Ga
7 | 35 Br

GALLIUM
GERMANIUM
ARSENIC
SELENIUM
BROMINE
KRYPTON
RUBIDIUM
STRONTIUM
YTTRIUM
ZIRCONIUM

ACROSS

36 Kr (1) 38 Sr (5) 33 As (8) 37 Rb (9) 34 Se (10)

Alkali Metal | Alkaline Earth | Transition Metal | Basic Metal | Metalloid | Nonmetal | Halogen | Noble Gas | Lanthanide | Actinide

DOWN

#	Element
1	32 Ge
3	37 Rb
4	36 Kr
6	38 Sr
8	33 As

FILL IN THE BLANKS

G _ L _ I _ M

G _ R _ A _ I _ M

A _ S _ N _ C

S _ L _ N _ U _

B _ O _ I _ E

K _ Y _ T _ N

R _ B _ D _ U _

S _ R _ N _ I _ M

Y _ T _ I _ M

Z _ R _ O _ I _ M

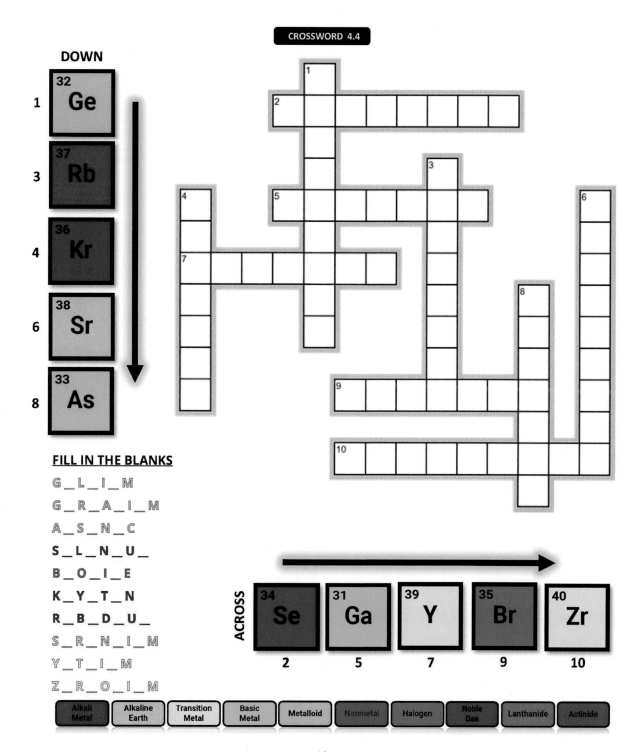

ACROSS

	2	5	7	9	10
	34 Se	31 Ga	39 Y	35 Br	40 Zr

Alkali Metal | Alkaline Earth | Transition Metal | Basic Metal | Metalloid | Nonmetal | Halogen | Noble Gas | Lanthanide | Actinide

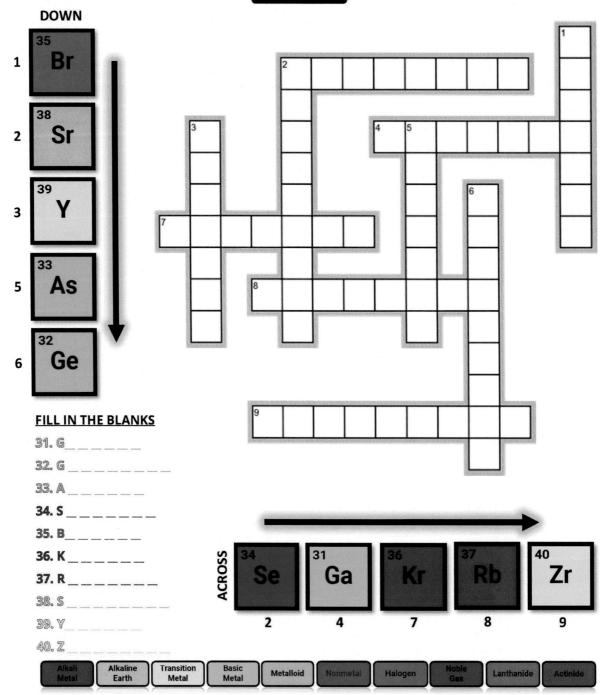

DOWN

1 — 35 **Br**

2 — 38 **Sr**

3 — 39 **Y**

5 — 33 **As**

6 — 32 **Ge**

FILL IN THE BLANKS

31. G _ _ _ _ _ _

32. G _ _ _ _ _ _ _ _ _

33. A _ _ _ _ _ _

34. S _ _ _ _ _ _ _

35. B _ _ _ _ _ _

36. K _ _ _ _ _ _ _

37. R _ _ _ _ _ _ _

38. S _ _ _ _ _ _ _

39. Y _ _ _ _ _ _

40. Z _ _ _ _ _ _ _ _

ACROSS

34 **Se** — 2

31 **Ga** — 4

36 **Kr** — 7

37 **Rb** — 8

40 **Zr** — 9

Alkali Metal | Alkaline Earth | Transition Metal | Basic Metal | Metalloid | Nonmetal | Halogen | Noble Gas | Lanthanide | Actinide

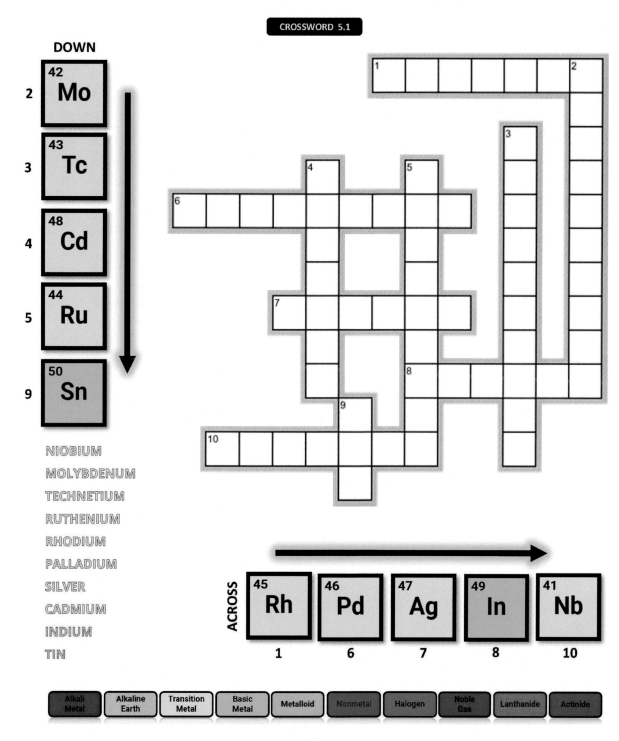

DOWN

2 — 42 Mo

3 — 43 Tc

4 — 48 Cd

5 — 44 Ru

9 — 50 Sn

NIOBIUM
MOLYBDENUM
TECHNETIUM
RUTHENIUM
RHODIUM
PALLADIUM
SILVER
CADMIUM
INDIUM
TIN

ACROSS

45 Rh — 1
46 Pd — 6
47 Ag — 7
49 In — 8
41 Nb — 10

Alkali Metal | Alkaline Earth | Transition Metal | Basic Metal | Metalloid | Nonmetal | Halogen | Noble Gas | Lanthanide | Actinide

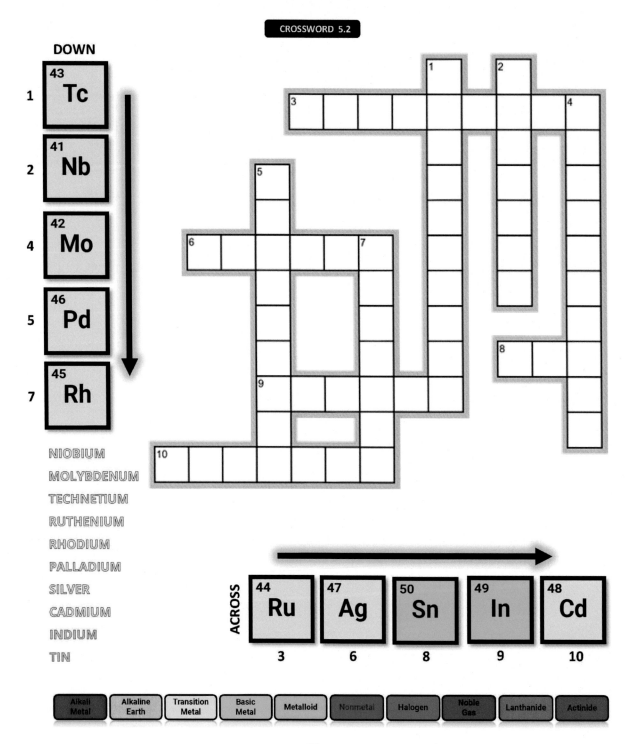

CROSSWORD 5.2

DOWN

1 — 43 Tc
2 — 41 Nb
4 — 42 Mo
5 — 46 Pd
7 — 45 Rh

NIOBIUM
MOLYBDENUM
TECHNETIUM
RUTHENIUM
RHODIUM
PALLADIUM
SILVER
CADMIUM
INDIUM
TIN

ACROSS

44 Ru — 3
47 Ag — 6
50 Sn — 8
49 In — 9
48 Cd — 10

Alkali Metal | Alkaline Earth | Transition Metal | Basic Metal | Metalloid | Nonmetal | Halogen | Noble Gas | Lanthanide | Actinide

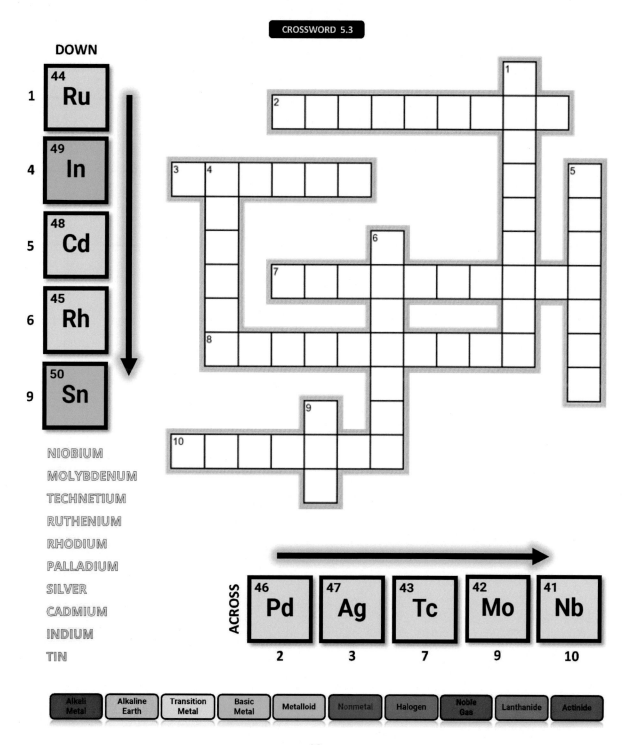

DOWN

1. 44 Ru
4. 49 In
5. 48 Cd
6. 45 Rh
9. 50 Sn

NIOBIUM
MOLYBDENUM
TECHNETIUM
RUTHENIUM
RHODIUM
PALLADIUM
SILVER
CADMIUM
INDIUM
TIN

ACROSS

46 Pd — 2
47 Ag — 3
43 Tc — 7
42 Mo — 9
41 Nb — 10

Alkali Metal | Alkaline Earth | Transition Metal | Basic Metal | Metalloid | Nonmetal | Halogen | Noble Gas | Lanthanide | Actinide

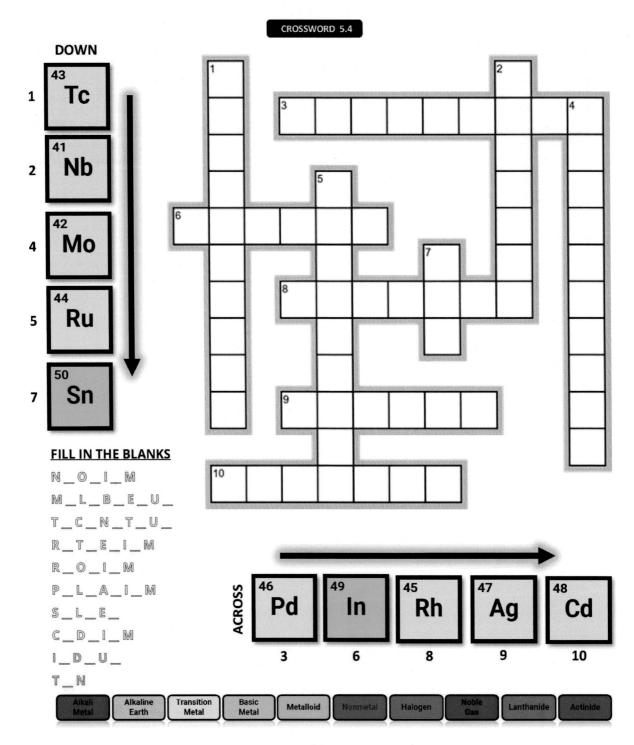

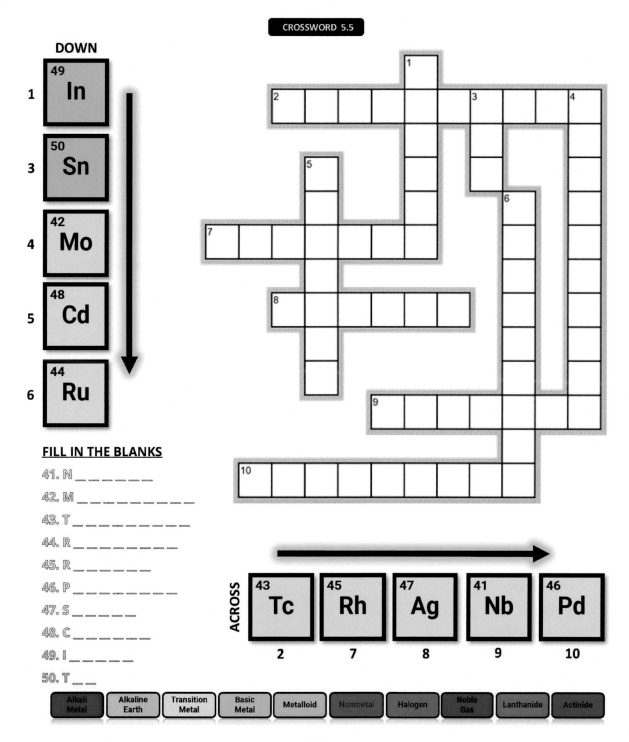

CROSSWORD 5.5

DOWN

1. ⁴⁹ In
3. ⁵⁰ Sn
4. ⁴² Mo
5. ⁴⁸ Cd
6. ⁴⁴ Ru

FILL IN THE BLANKS

41. N _ _ _ _ _ _ _
42. M _ _ _ _ _ _ _ _ _
43. T _ _ _ _ _ _ _ _
44. R _ _ _ _ _ _ _ _
45. R _ _ _ _ _ _
46. P _ _ _ _ _ _ _ _
47. S _ _ _ _ _ _
48. C _ _ _ _ _ _
49. I _ _ _ _ _ _
50. T _ _ _

ACROSS

2	7	8	9	10
⁴³ Tc	⁴⁵ Rh	⁴⁷ Ag	⁴¹ Nb	⁴⁶ Pd

Alkali Metal | Alkaline Earth | Transition Metal | Basic Metal | Metalloid | Nonmetal | Halogen | Noble Gas | Lanthanide | Actinide

25

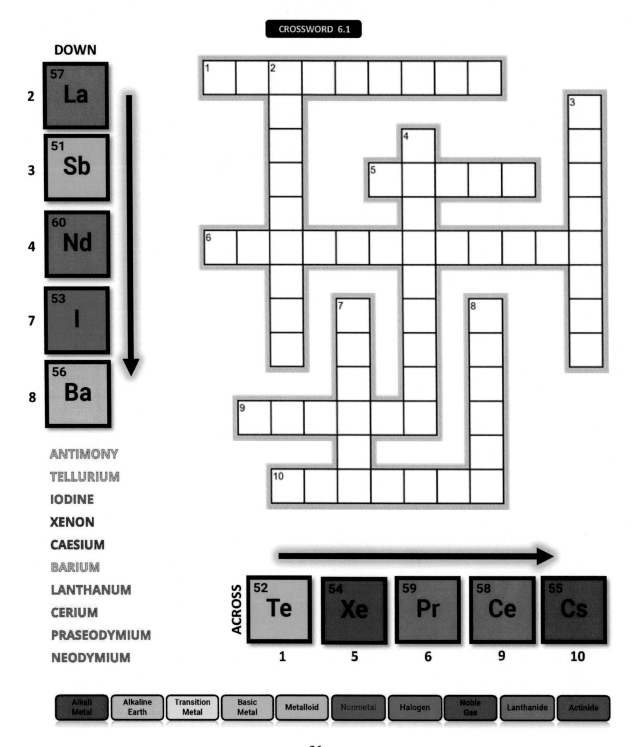

CROSSWORD 6.1

DOWN

2 — 57 La
3 — 51 Sb
4 — 60 Nd
7 — 53 I
8 — 56 Ba

ANTIMONY
TELLURIUM
IODINE
XENON
CAESIUM
BARIUM
LANTHANUM
CERIUM
PRASEODYMIUM
NEODYMIUM

ACROSS

52 Te — 1
54 Xe — 5
59 Pr — 6
58 Ce — 9
55 Cs — 10

Alkali Metal | Alkaline Earth | Transition Metal | Basic Metal | Metalloid | Nonmetal | Halogen | Noble Gas | Lanthanide | Actinide

26

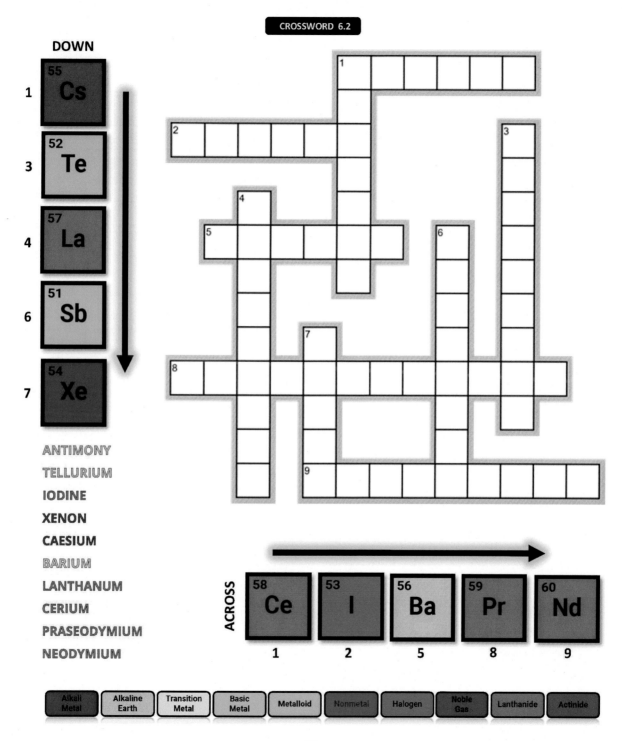

CROSSWORD 6.2

DOWN

1 | **Cs** 55

3 | **Te** 52

4 | **La** 57

6 | **Sb** 51

7 | **Xe** 54

ANTIMONY
TELLURIUM
IODINE
XENON
CAESIUM
BARIUM
LANTHANUM
CERIUM
PRASEODYMIUM
NEODYMIUM

ACROSS

Ce 58 — 1
I 53 — 2
Ba 56 — 5
Pr 59 — 8
Nd 60 — 9

Alkali Metal | Alkaline Earth | Transition Metal | Basic Metal | Metalloid | Nonmetal | Halogen | Noble Gas | Lanthanide | Actinide

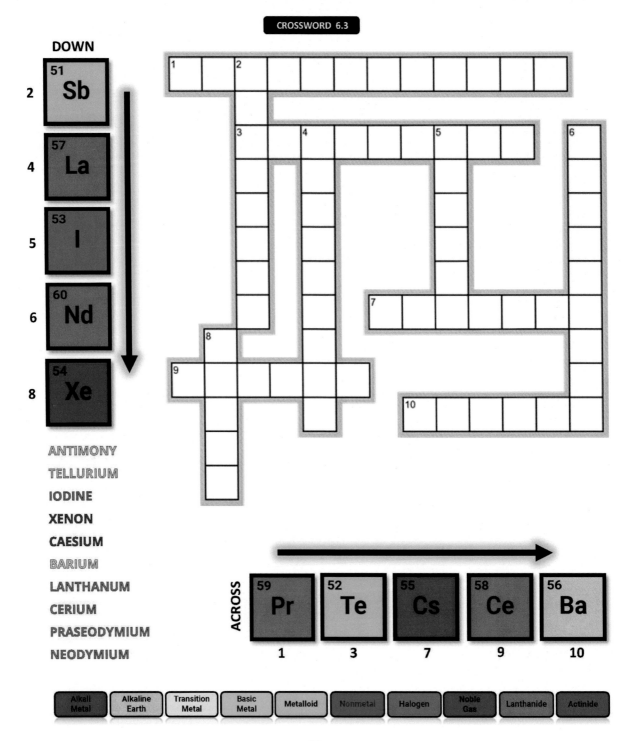

CROSSWORD 6.3

DOWN

51 **Sb** — 2

57 **La** — 4

53 **I** — 5

60 **Nd** — 6

54 **Xe** — 8

ANTIMONY
TELLURIUM
IODINE
XENON
CAESIUM
BARIUM
LANTHANUM
CERIUM
PRASEODYMIUM
NEODYMIUM

ACROSS

59 **Pr** — 1
52 **Te** — 3
55 **Cs** — 7
58 **Ce** — 9
56 **Ba** — 10

| Alkali Metal | Alkaline Earth | Transition Metal | Basic Metal | Metalloid | Nonmetal | Halogen | Noble Gas | Lanthanide | Actinide |

CROSSWORD 6.4

DOWN

2	**57 La**
3	**51 Sb**
4	**56 Ba**
6	**58 Ce**
8	**53 I**

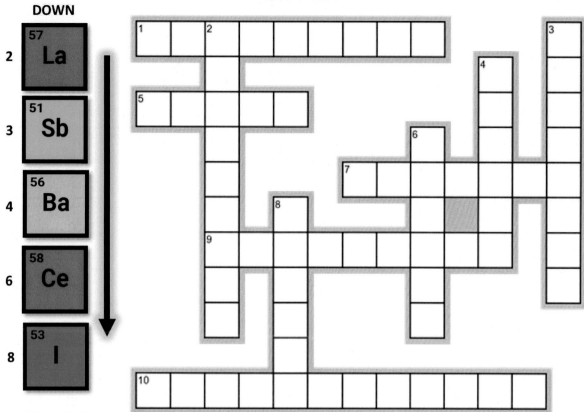

FILL IN THE BLANKS

A _ T _ M _ N _

T _ L _ U _ I _ M

I _ D _ N _

X _ N _ N

C _ E _ I _ M

B _ R _ U _

L _ N _ H _ N _ M

C _ R _ U _

P _ A _ E _ D _ M _ U _

N _ O _ Y _ I _ M

ACROSS

52 Te	54 Xe	55 Cs	60 Nd	59 Pr
1	5	7	9	10

Alkali Metal	Alkaline Earth	Transition Metal	Basic Metal	Metalloid	Nonmetal	Halogen	Noble Gas	Lanthanide	Actinide

CROSSWORD 6.5

DOWN

2. **51 Sb**
3. **58 Ce**
4. **59 Pr**
5. **52 Te**
6. **54 Xe**

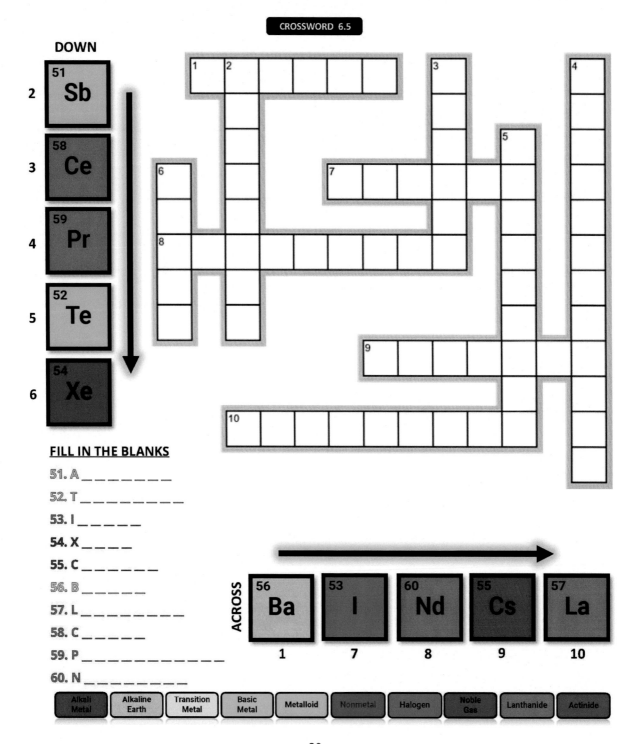

FILL IN THE BLANKS

51. A _ _ _ _ _ _ _ _
52. T _ _ _ _ _ _ _ _
53. I _ _ _ _ _ _
54. X _ _ _ _ _
55. C _ _ _ _ _ _ _
56. B _ _ _ _ _ _
57. L _ _ _ _ _ _ _ _ _
58. C _ _ _ _ _ _
59. P _ _ _ _ _ _ _ _ _ _ _ _
60. N _ _ _ _ _ _ _ _ _

ACROSS

1. **56 Ba**
7. **53 I**
8. **60 Nd**
9. **55 Cs**
10. **57 La**

Alkali Metal	Alkaline Earth	Transition Metal	Basic Metal	Metalloid	Nonmetal	Halogen	Noble Gas	Lanthanide	Actinide

CROSSWORD 7.1

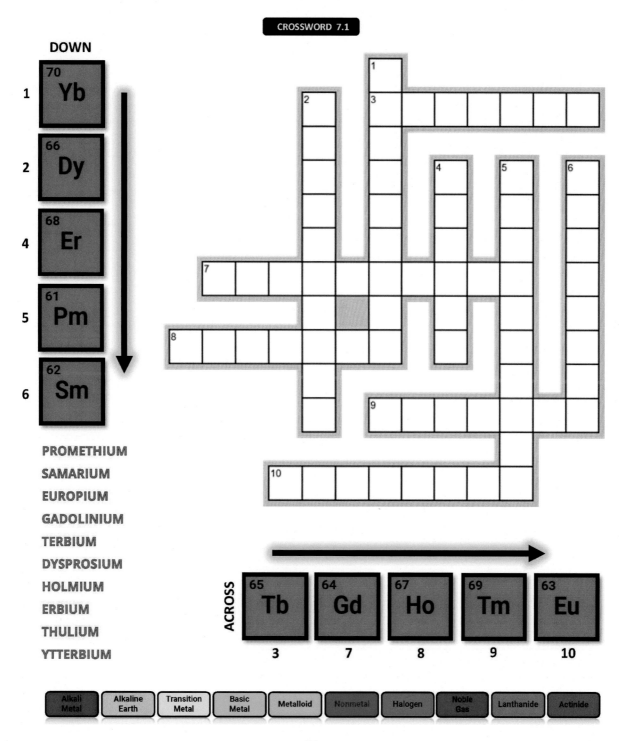

31

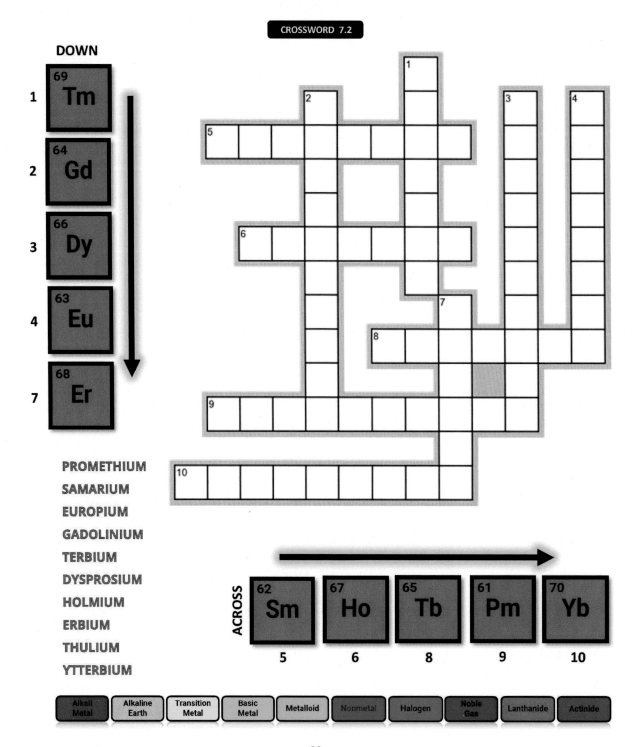

CROSSWORD 7.2

DOWN

1 | 69 Tm |
2 | 64 Gd |
3 | 66 Dy |
4 | 63 Eu |
7 | 68 Er |

PROMETHIUM
SAMARIUM
EUROPIUM
GADOLINIUM
TERBIUM
DYSPROSIUM
HOLMIUM
ERBIUM
THULIUM
YTTERBIUM

ACROSS

| 62 Sm | 67 Ho | 65 Tb | 61 Pm | 70 Yb |
| 5 | 6 | 8 | 9 | 10 |

Alkali Metal | Alkaline Earth | Transition Metal | Basic Metal | Metalloid | Nonmetal | Halogen | Noble Gas | Lanthanide | Actinide

32

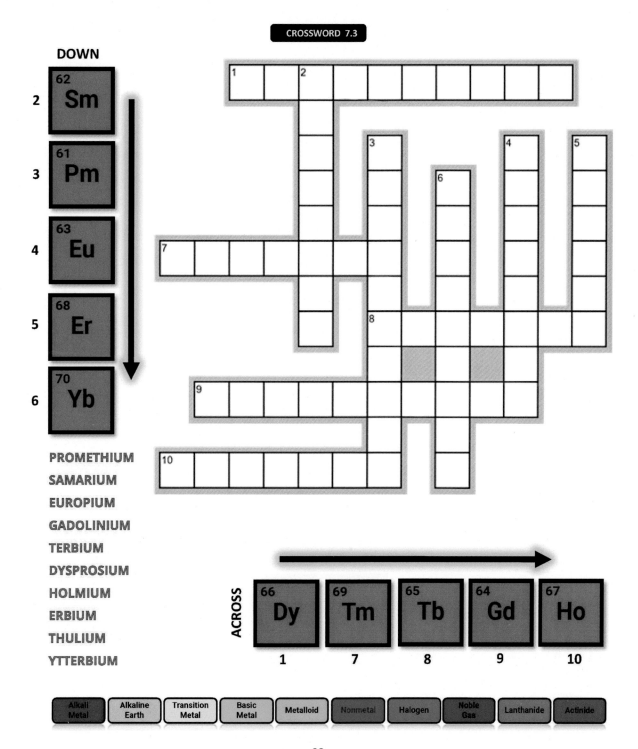

CROSSWORD 7.3

DOWN

2 Sm (62)
3 Pm (61)
4 Eu (63)
5 Er (68)
6 Yb (70)

PROMETHIUM
SAMARIUM
EUROPIUM
GADOLINIUM
TERBIUM
DYSPROSIUM
HOLMIUM
ERBIUM
THULIUM
YTTERBIUM

ACROSS
1 Dy (66)
7 Tm (69)
8 Tb (65)
9 Gd (64)
10 Ho (67)

Alkali Metal | Alkaline Earth | Transition Metal | Basic Metal | Metalloid | Nonmetal | Halogen | Noble Gas | Lanthanide | Actinide

33

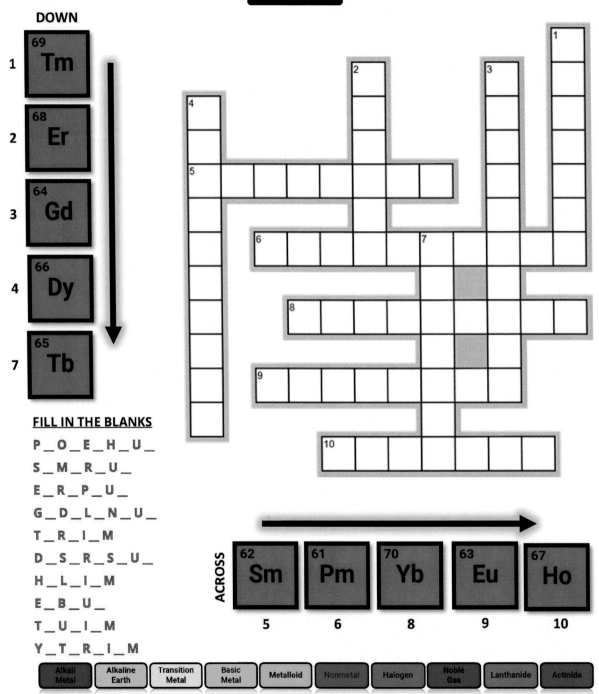

DOWN

1 | 69 Tm
2 | 68 Er
3 | 64 Gd
4 | 66 Dy
7 | 65 Tb

FILL IN THE BLANKS

P _ O _ E _ H _ U _

S _ M _ R _ U _

E _ R _ P _ U _

G _ D _ L _ N _ U _

T _ R _ I _ M

D _ S _ R _ S _ U _

H _ L _ I _ M

E _ B _ U _

T _ U _ I _ M

Y _ T _ R _ I _ M

ACROSS

62 Sm	61 Pm	70 Yb	63 Eu	67 Ho
5	6	8	9	10

Alkali Metal | Alkaline Earth | Transition Metal | Basic Metal | Metalloid | Nonmetal | Halogen | Noble Gas | Lanthanide | Actinide

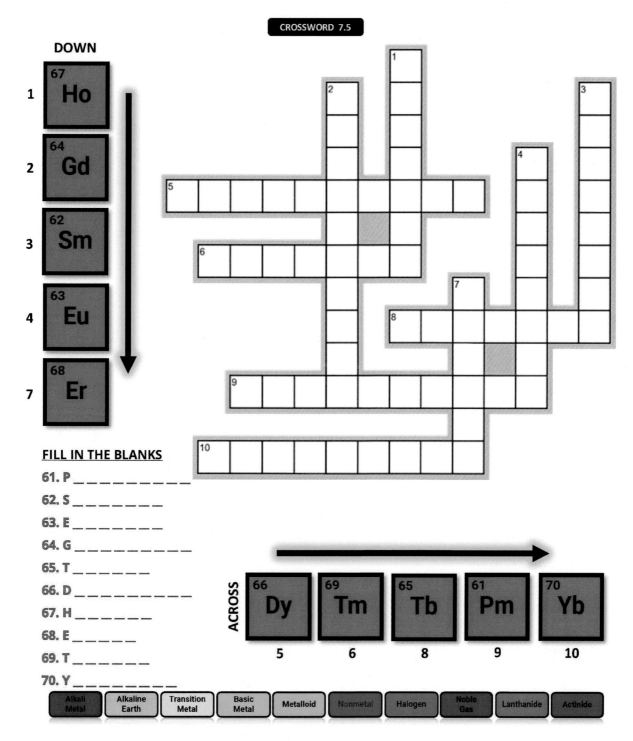

DOWN

1 ⁶⁷ Ho

2 ⁶⁴ Gd

3 ⁶² Sm

4 ⁶³ Eu

7 ⁶⁸ Er

FILL IN THE BLANKS

61. P _ _ _ _ _ _ _ _ _ _

62. S _ _ _ _ _ _ _ _

63. E _ _ _ _ _ _ _ _

64. G _ _ _ _ _ _ _ _ _

65. T _ _ _ _ _ _ _

66. D _ _ _ _ _ _ _ _ _ _

67. H _ _ _ _ _ _ _

68. E _ _ _ _ _ _

69. T _ _ _ _ _ _ _

70. Y _ _ _ _ _ _ _ _

ACROSS

⁶⁶ Dy 5

⁶⁹ Tm 6

⁶⁵ Tb 8

⁶¹ Pm 9

⁷⁰ Yb 10

Alkali Metal	Alkaline Earth	Transition Metal	Basic Metal	Metalloid	Nonmetal	Halogen	Noble Gas	Lanthanide	Actinide

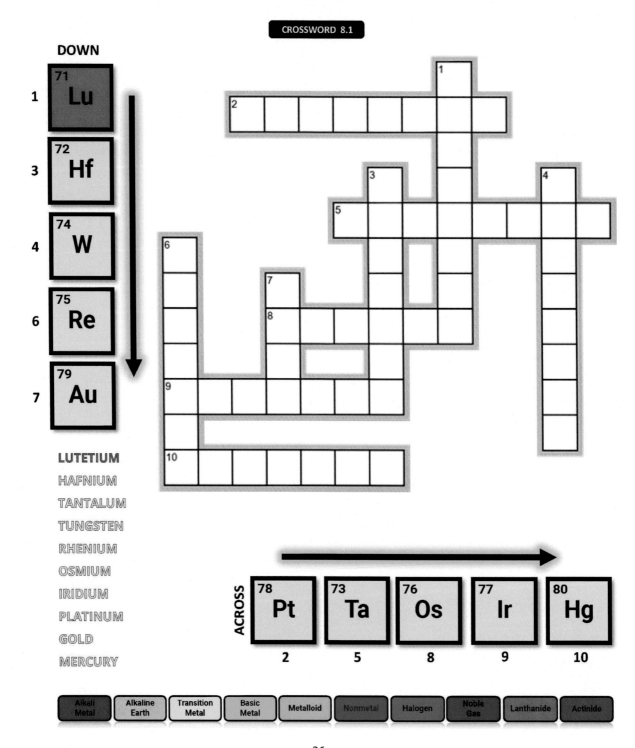

CROSSWORD 8.1

DOWN

#	Element
1	71 Lu
3	72 Hf
4	74 W
6	75 Re
7	79 Au

LUTETIUM
HAFNIUM
TANTALUM
TUNGSTEN
RHENIUM
OSMIUM
IRIDIUM
PLATINUM
GOLD
MERCURY

ACROSS

Element	#
78 Pt	2
73 Ta	5
76 Os	8
77 Ir	9
80 Hg	10

Alkali Metal | Alkaline Earth | Transition Metal | Basic Metal | Metalloid | Nonmetal | Halogen | Noble Gas | Lanthanide | Actinide

CROSSWORD 8.2

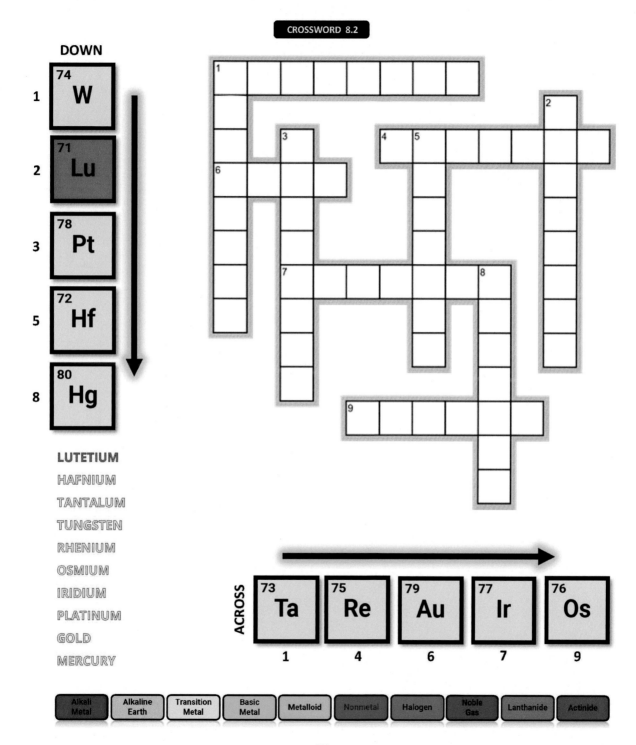

DOWN

1 74 W

2 71 Lu

3 78 Pt

5 72 Hf

8 80 Hg

LUTETIUM
HAFNIUM
TANTALUM
TUNGSTEN
RHENIUM
OSMIUM
IRIDIUM
PLATINUM
GOLD
MERCURY

ACROSS

73 Ta   75 Re   79 Au   77 Ir   76 Os
  1   4   6   7   9

Alkali Metal | Alkaline Earth | Transition Metal | Basic Metal | Metalloid | Nonmetal | Halogen | Noble Gas | Lanthanide | Actinide

37

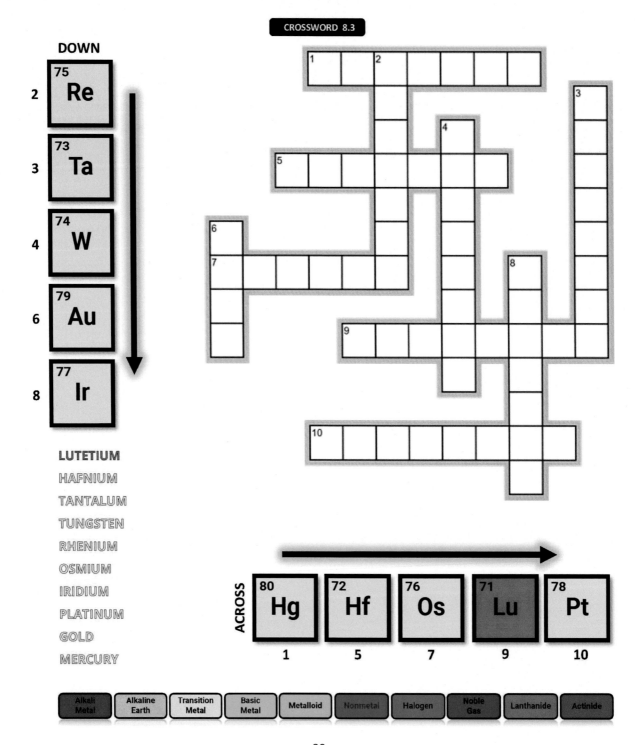

CROSSWORD 8.3

DOWN

2 | 75 Re
3 | 73 Ta
4 | 74 W
6 | 79 Au
8 | 77 Ir

LUTETIUM
HAFNIUM
TANTALUM
TUNGSTEN
RHENIUM
OSMIUM
IRIDIUM
PLATINUM
GOLD
MERCURY

ACROSS

80 Hg — 1
72 Hf — 5
76 Os — 7
71 Lu — 9
78 Pt — 10

Alkali Metal | Alkaline Earth | Transition Metal | Basic Metal | Metalloid | Nonmetal | Halogen | Noble Gas | Lanthanide | Actinide

38

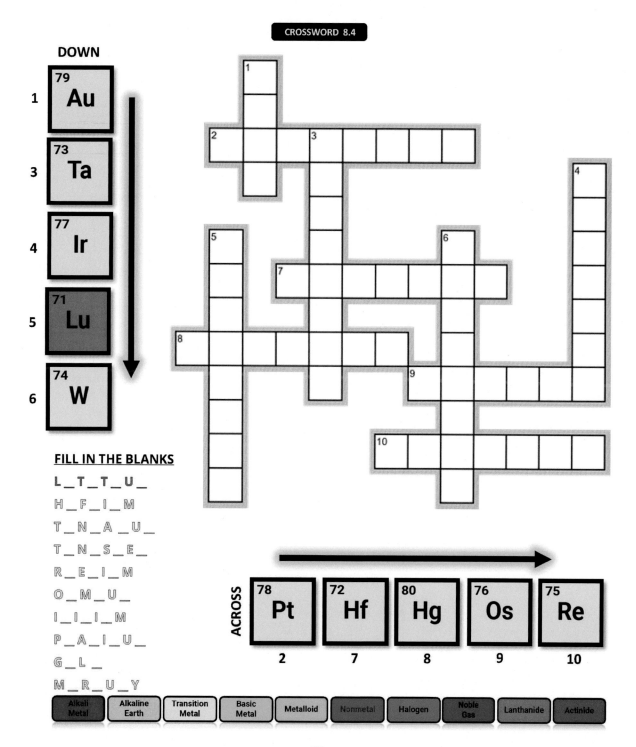

DOWN

1. Au 79
3. Ta 73
4. Ir 77
5. Lu 71
6. W 74

FILL IN THE BLANKS

L _ T _ T _ U _

H _ F _ I _ M

T _ N _ A _ U _

T _ N _ S _ E _

R _ E _ I _ M

O _ M _ U _

I _ I _ I _ M

P _ A _ I _ U _

G _ L _

M _ R _ U _ Y

ACROSS

2. Pt 78
7. Hf 72
8. Hg 80
9. Os 76
10. Re 75

Alkali Metal | Alkaline Earth | Transition Metal | Basic Metal | Metalloid | Nonmetal | Halogen | Noble Gas | Lanthanide | Actinide

CROSSWORD 8.5

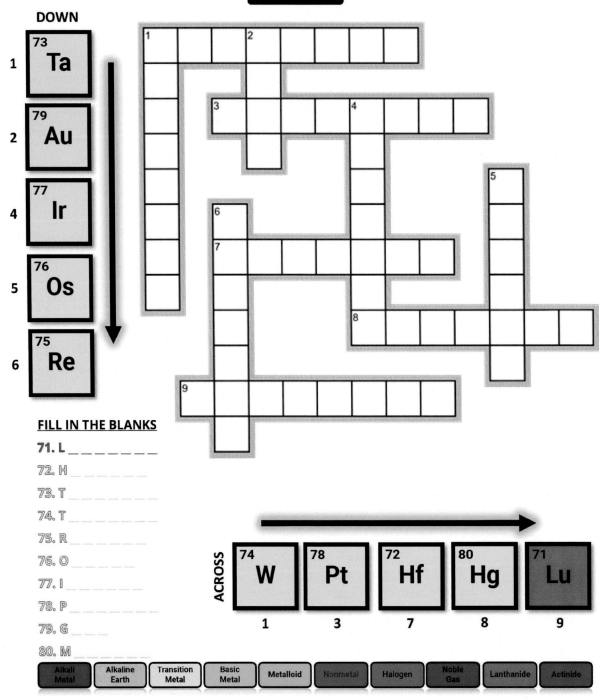

DOWN

#	Symbol	Atomic Number
1	Ta	73
2	Au	79
4	Ir	77
5	Os	76
6	Re	75

FILL IN THE BLANKS

71. L _ _ _ _ _ _ _ _
72. H _ _ _ _ _ _
73. T _ _ _ _ _ _ _
74. T _ _ _ _ _ _ _
75. R _ _ _ _ _ _
76. O _ _ _ _ _
77. I _ _ _ _ _ _
78. P _ _ _ _ _ _
79. G _ _ _
80. M _ _ _ _ _ _

ACROSS

74 W	78 Pt	72 Hf	80 Hg	71 Lu
1	3	7	8	9

Alkali Metal	Alkaline Earth	Transition Metal	Basic Metal	Metalloid	Nonmetal	Halogen	Noble Gas	Lanthanide	Actinide

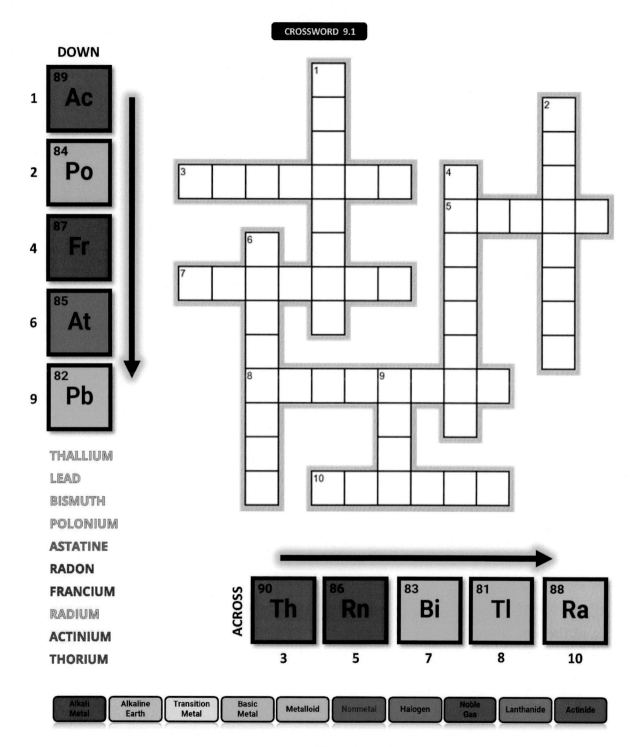

DOWN

1 89 Ac

2 84 Po

4 87 Fr

6 85 At

9 82 Pb

THALLIUM
LEAD
BISMUTH
POLONIUM
ASTATINE
RADON
FRANCIUM
RADIUM
ACTINIUM
THORIUM

ACROSS

90 Th — 3
86 Rn — 5
83 Bi — 7
81 Tl — 8
88 Ra — 10

Alkali Metal | Alkaline Earth | Transition Metal | Basic Metal | Metalloid | Nonmetal | Halogen | Noble Gas | Lanthanide | Actinide

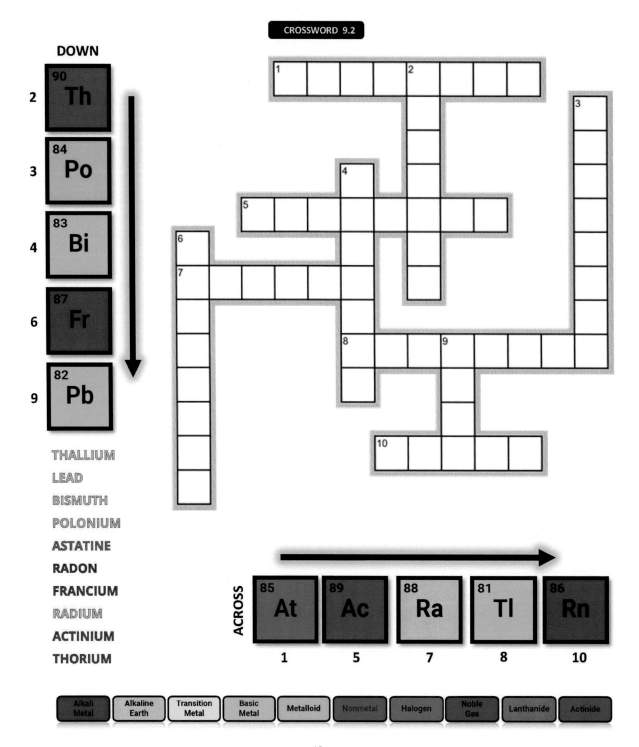

CROSSWORD 9.2

DOWN

2	90 Th
3	84 Po
4	83 Bi
6	87 Fr
9	82 Pb

THALLIUM
LEAD
BISMUTH
POLONIUM
ASTATINE
RADON
FRANCIUM
RADIUM
ACTINIUM
THORIUM

ACROSS

| | 85 At | 89 Ac | 88 Ra | 81 Tl | 86 Rn |
| | 1 | 5 | 7 | 8 | 10 |

| Alkali Metal | Alkaline Earth | Transition Metal | Basic Metal | Metalloid | Nonmetal | Halogen | Noble Gas | Lanthanide | Actinide |

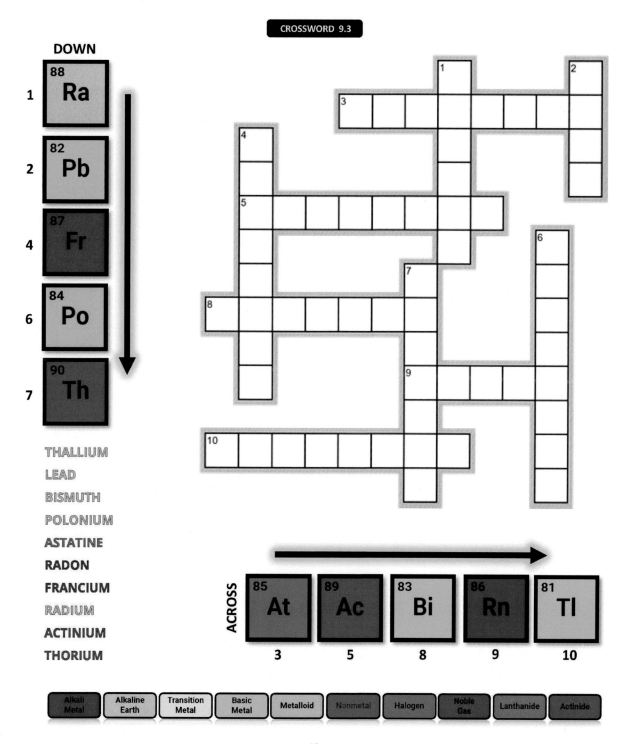

DOWN

1. 88 Ra
2. 82 Pb
4. 87 Fr
6. 84 Po
7. 90 Th

THALLIUM
LEAD
BISMUTH
POLONIUM
ASTATINE
RADON
FRANCIUM
RADIUM
ACTINIUM
THORIUM

ACROSS

3. 85 At
5. 89 Ac
8. 83 Bi
9. 86 Rn
10. 81 Tl

Alkali Metal | Alkaline Earth | Transition Metal | Basic Metal | Metalloid | Nonmetal | Halogen | Noble Gas | Lanthanide | Actinide

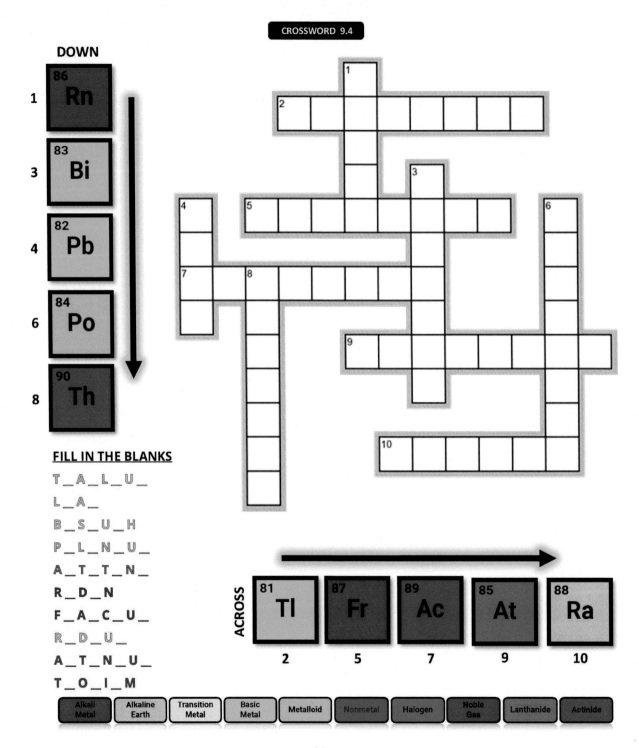

CROSSWORD 9.4

DOWN

#	Z	Symbol
1	86	Rn
3	83	Bi
4	82	Pb
6	84	Po
8	90	Th

FILL IN THE BLANKS

T _ A _ L _ U _

L _ A _

B _ S _ U _ H

P _ L _ N _ U _

A _ T _ T _ N _

R _ D _ N

F _ A _ C _ U _

R _ D _ U _

A _ T _ N _ U _

T _ O _ I _ M

ACROSS

#	Z	Symbol	Clue #
	81	Tl	2
	87	Fr	5
	89	Ac	7
	85	At	9
	88	Ra	10

Alkali Metal | Alkaline Earth | Transition Metal | Basic Metal | Metalloid | Nonmetal | Halogen | Noble Gas | Lanthanide | Actinide

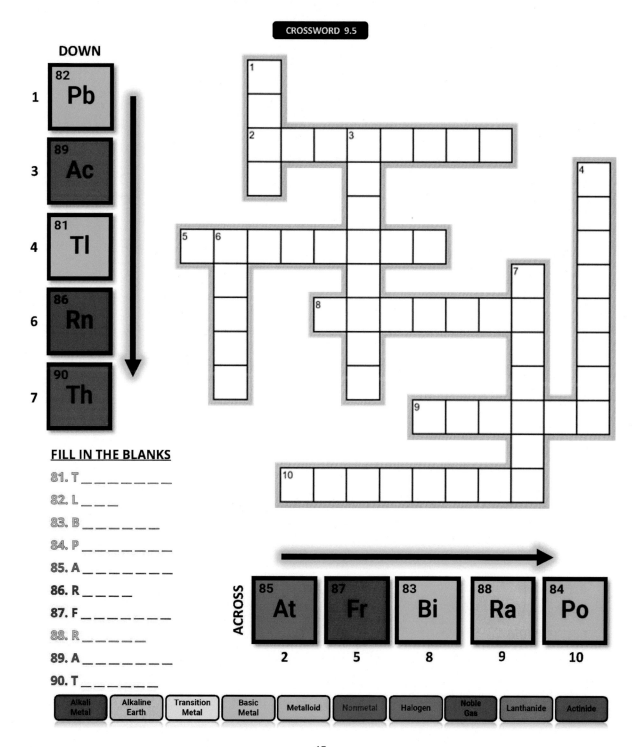

DOWN

1. 82 Pb

3. 89 Ac

4. 81 Tl

6. 86 Rn

7. 90 Th

FILL IN THE BLANKS

81. T _ _ _ _ _ _ _ _

82. L _ _ _ _

83. B _ _ _ _ _ _

84. P _ _ _ _ _ _ _

85. A _ _ _ _ _ _ _

86. R _ _ _ _

87. F _ _ _ _ _ _ _ _

88. R _ _ _ _ _

89. A _ _ _ _ _ _ _ _

90. T _ _ _ _ _ _

ACROSS

85 At — 2
87 Fr — 5
83 Bi — 8
88 Ra — 9
84 Po — 10

Alkali Metal | Alkaline Earth | Transition Metal | Basic Metal | Metalloid | Nonmetal | Halogen | Noble Gas | Lanthanide | Actinide

DOWN

1 | 91 Pa
2 | 92 U
3 | 97 Bk
4 | 95 Am
5 | 99 Es

PROTACTINIUM
URANIUM
NEPTUNIUM
PLUTONIUM
AMERICIUM
CURIUM
BERKELIUM
CALIFORNIUM
EINSTEINIUM
FERMIUM

ACROSS

1 | 94 Pu
6 | 96 Cm
7 | 98 Cf
8 | 100 Fm
9 | 93 Np

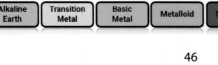

Alkali Metal | Alkaline Earth | Transition Metal | Basic Metal | Metalloid | Nonmetal | Halogen | Noble Gas | Lanthanide | Actinide

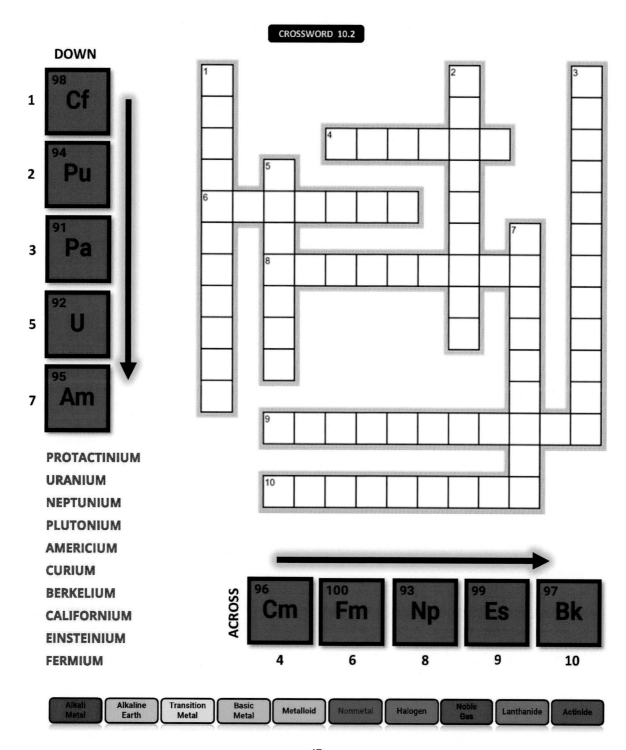

CROSSWORD 10.2

DOWN

1 — 98 Cf
2 — 94 Pu
3 — 91 Pa
5 — 92 U
7 — 95 Am

PROTACTINIUM
URANIUM
NEPTUNIUM
PLUTONIUM
AMERICIUM
CURIUM
BERKELIUM
CALIFORNIUM
EINSTEINIUM
FERMIUM

ACROSS

4 — 96 Cm
6 — 100 Fm
8 — 93 Np
9 — 99 Es
10 — 97 Bk

Alkali Metal | Alkaline Earth | Transition Metal | Basic Metal | Metalloid | Nonmetal | Halogen | Noble Gas | Lanthanide | Actinide

CROSSWORD 10.3

DOWN

#	Z	Symbol
2	92	U
3	94	Pu
4	95	Am
6	97	Bk
8	93	Np

PROTACTINIUM
URANIUM
NEPTUNIUM
PLUTONIUM
AMERICIUM
CURIUM
BERKELIUM
CALIFORNIUM
EINSTEINIUM
FERMIUM

ACROSS

#	Z	Symbol
1	91	Pa
5	98	Cf
7	99	Es
9	96	Cm
10	100	Fm

Alkali Metal | Alkaline Earth | Transition Metal | Basic Metal | Metalloid | Nonmetal | Halogen | Noble Gas | Lanthanide | Actinide

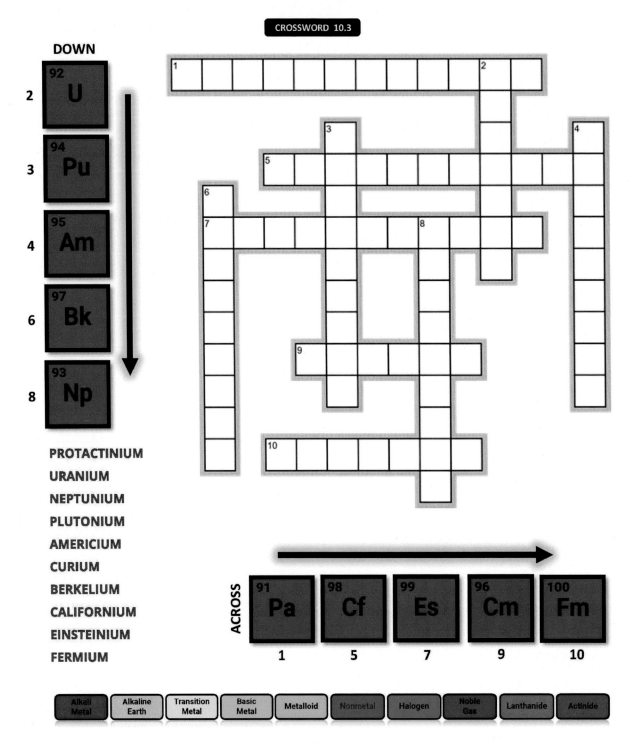

DOWN

#	Element
2	96 Cm
3	93 Np
4	95 Am
5	92 U
7	100 Fm

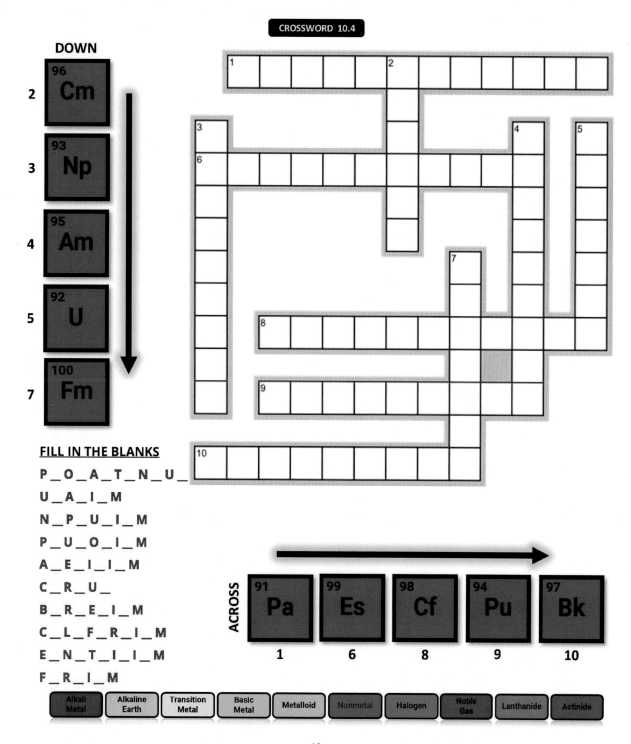

FILL IN THE BLANKS

P _ O _ A _ T _ N _ U _

U _ A _ I _ M

N _ P _ U _ I _ M

P _ U _ O _ I _ M

A _ E _ I _ I _ M

C _ R _ U _

B _ R _ E _ I _ M

C _ L _ F _ R _ I _ M

E _ N _ T _ I _ I _ M

F _ _ R _ I _ M

ACROSS

#	Element
1	91 Pa
6	99 Es
8	98 Cf
9	94 Pu
10	97 Bk

Alkali Metal	Alkaline Earth	Transition Metal	Basic Metal	Metalloid	Nonmetal	Halogen	Noble Gas	Lanthanide	Actinide

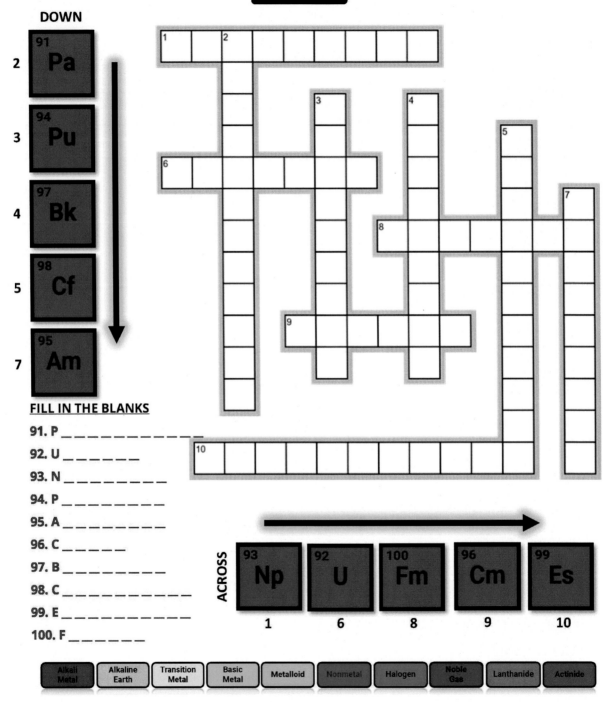

CROSSWORD 10.5

DOWN

2. 91 Pa
3. 94 Pu
4. 97 Bk
5. 98 Cf
7. 95 Am

FILL IN THE BLANKS

91. P _ _ _ _ _ _ _ _ _ _ _ _
92. U _ _ _ _ _ _ _
93. N _ _ _ _ _ _ _ _ _
94. P _ _ _ _ _ _ _ _
95. A _ _ _ _ _ _ _ _
96. C _ _ _ _ _
97. B _ _ _ _ _ _ _ _
98. C _ _ _ _ _ _ _ _ _ _
99. E _ _ _ _ _ _ _ _ _ _
100. F _ _ _ _ _ _

ACROSS

1. 93 Np
6. 92 U
8. 100 Fm
9. 96 Cm
10. 99 Es

Alkali Metal | Alkaline Earth | Transition Metal | Basic Metal | Metalloid | Nonmetal | Halogen | Noble Gas | Lanthanide | Actinide

CROSSWORD 11.1

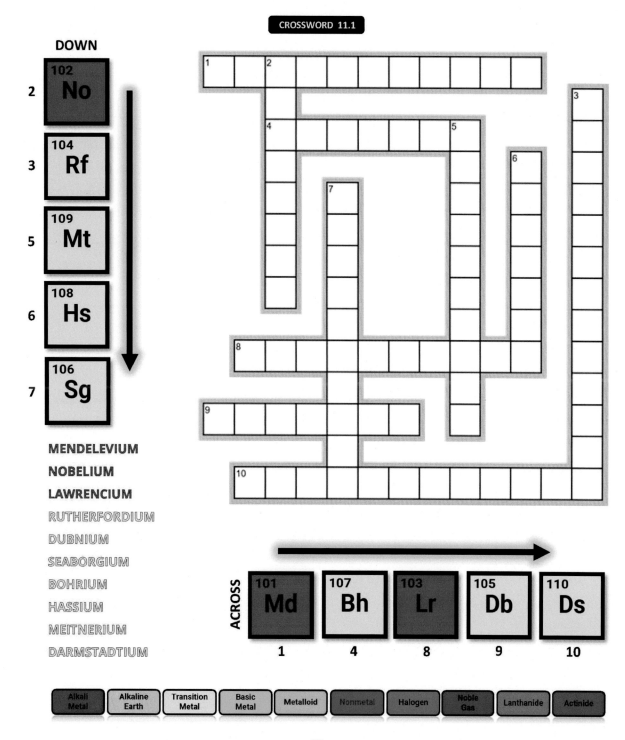

DOWN

2. No — 102
3. Rf — 104
5. Mt — 109
6. Hs — 108
7. Sg — 106

MENDELEVIUM
NOBELIUM
LAWRENCIUM
RUTHERFORDIUM
DUBNIUM
SEABORGIUM
BOHRIUM
HASSIUM
MEITNERIUM
DARMSTADTIUM

ACROSS

Md — 101 — 1
Bh — 107 — 4
Lr — 103 — 8
Db — 105 — 9
Ds — 110 — 10

Alkali Metal | Alkaline Earth | Transition Metal | Basic Metal | Metalloid | Nonmetal | Halogen | Noble Gas | Lanthanide | Actinide

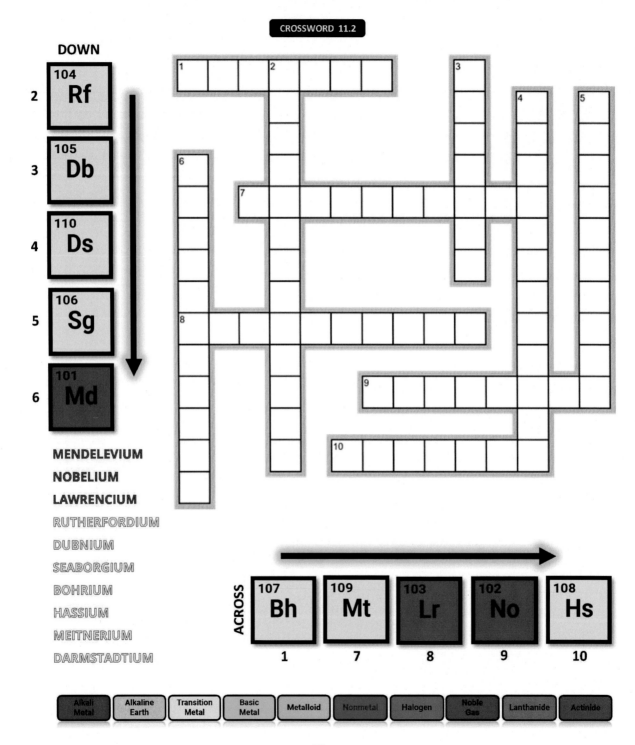

CROSSWORD 11.2

DOWN

#	Symbol	Number
2	Rf	104
3	Db	105
4	Ds	110
5	Sg	106
6	Md	101

MENDELEVIUM
NOBELIUM
LAWRENCIUM
RUTHERFORDIUM
DUBNIUM
SEABORGIUM
BOHRIUM
HASSIUM
MEITNERIUM
DARMSTADTIUM

ACROSS

#	Symbol	Number
1	Bh	107
7	Mt	109
8	Lr	103
9	No	102
10	Hs	108

Alkali Metal | Alkaline Earth | Transition Metal | Basic Metal | Metalloid | Nonmetal | Halogen | Noble Gas | Lanthanide | Actinide

52

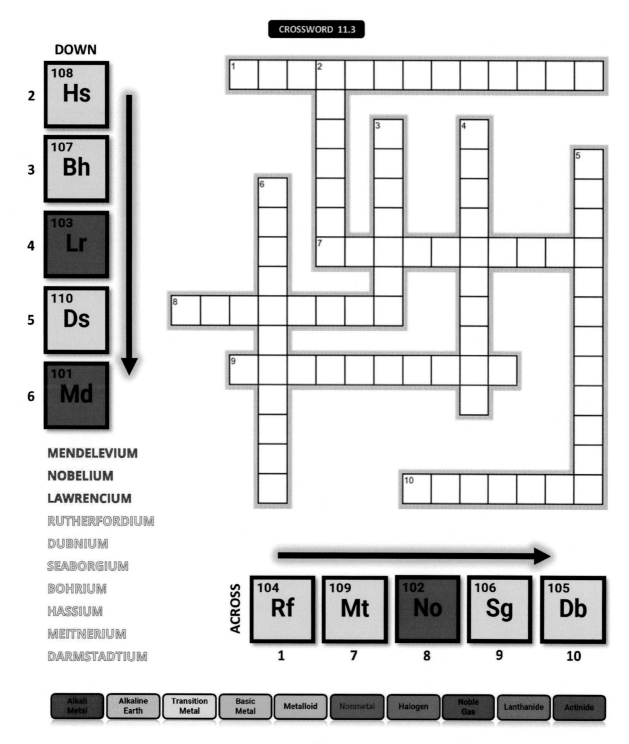

CROSSWORD 11.3

DOWN

2 — 108 Hs
3 — 107 Bh
4 — 103 Lr
5 — 110 Ds
6 — 101 Md

MENDELEVIUM
NOBELIUM
LAWRENCIUM
RUTHERFORDIUM
DUBNIUM
SEABORGIUM
BOHRIUM
HASSIUM
MEITNERIUM
DARMSTADTIUM

ACROSS

1 — 104 Rf
7 — 109 Mt
8 — 102 No
9 — 106 Sg
10 — 105 Db

Alkali Metal | Alkaline Earth | Transition Metal | Basic Metal | Metalloid | Nonmetal | Halogen | Noble Gas | Lanthanide | Actinide

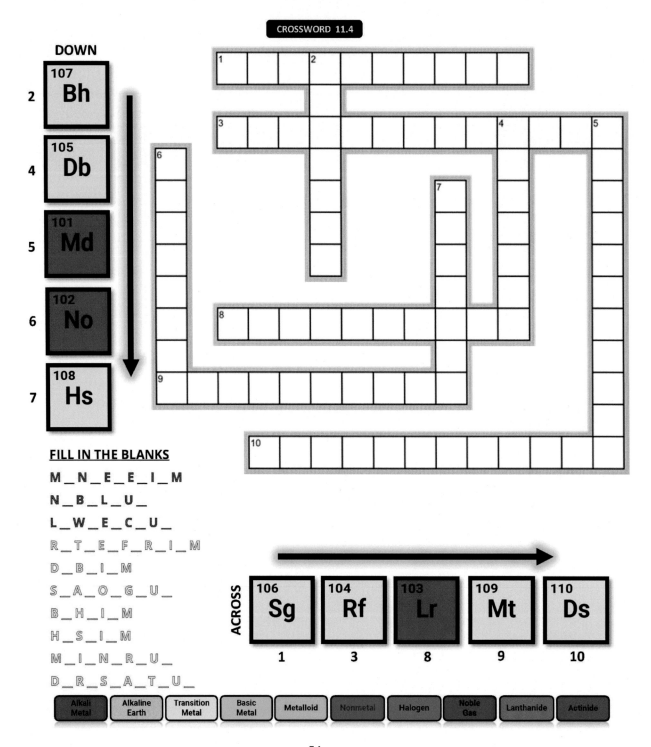

CROSSWORD 11.4

DOWN

2	107 **Bh**	
4	105 **Db**	
5	101 **Md**	
6	102 **No**	
7	108 **Hs**	

FILL IN THE BLANKS

M _ N _ E _ E _ I _ M

N _ B _ L _ U _

L _ W _ E _ C _ U _

R _ T _ E _ F _ R _ I _ M

D _ B _ I _ M

S _ A _ O _ G _ U _

B _ H _ I _ M

H _ S _ I _ M

M _ I _ N _ R _ U _

D _ R _ S _ A _ T _ U _

ACROSS

106 **Sg**	104 **Rf**	103 **Lr**	109 **Mt**	110 **Ds**
1	3	8	9	10

Alkali Metal	Alkaline Earth	Transition Metal	Basic Metal	Metalloid	Nonmetal	Halogen	Noble Gas	Lanthanide	Actinide

DOWN

#		
1	101	Md
2	104	Rf
3	106	Sg
4	105	Db
6	107	Bh

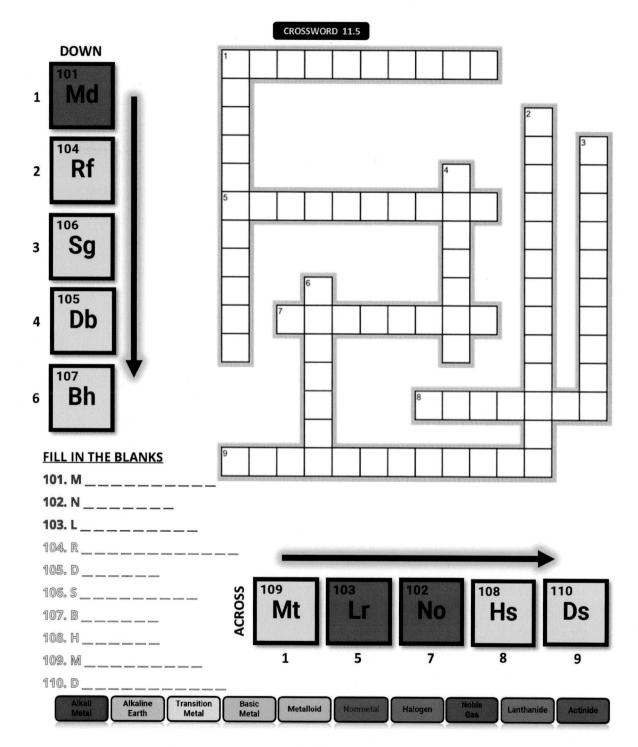

FILL IN THE BLANKS

101. M _ _ _ _ _ _ _ _ _ _ _

102. N _ _ _ _ _ _ _ _

103. L _ _ _ _ _ _ _ _ _ _

104. R _ _ _ _ _ _ _ _ _ _ _ _ _

105. D _ _ _ _ _ _ _

106. S _ _ _ _ _ _ _ _ _ _

107. B _ _ _ _ _ _ _

108. H _ _ _ _ _ _ _

109. M _ _ _ _ _ _ _ _ _ _

110. D _ _ _ _ _ _ _ _ _ _ _ _

ACROSS

109 Mt	103 Lr	102 No	108 Hs	110 Ds
1	5	7	8	9

Alkali Metal	Alkaline Earth	Transition Metal	Basic Metal	Metalloid	Nonmetal	Halogen	Noble Gas	Lanthanide	Actinide

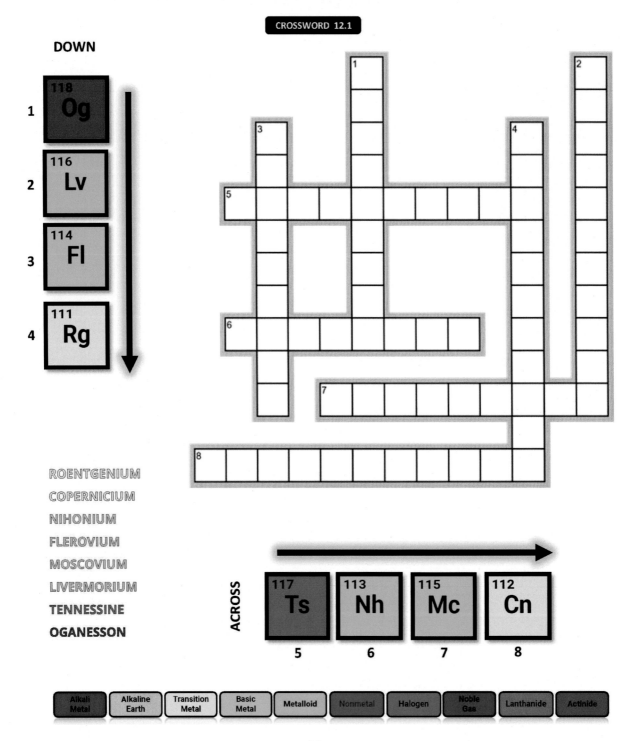

CROSSWORD 12.1

DOWN

1. Og 118
2. Lv 116
3. Fl 114
4. Rg 111

ROENTGENIUM
COPERNICIUM
NIHONIUM
FLEROVIUM
MOSCOVIUM
LIVERMORIUM
TENNESSINE
OGANESSON

ACROSS

5. Ts 117
6. Nh 113
7. Mc 115
8. Cn 112

Alkali Metal | Alkaline Earth | Transition Metal | Basic Metal | Metalloid | Nonmetal | Halogen | Noble Gas | Lanthanide | Actinide

56

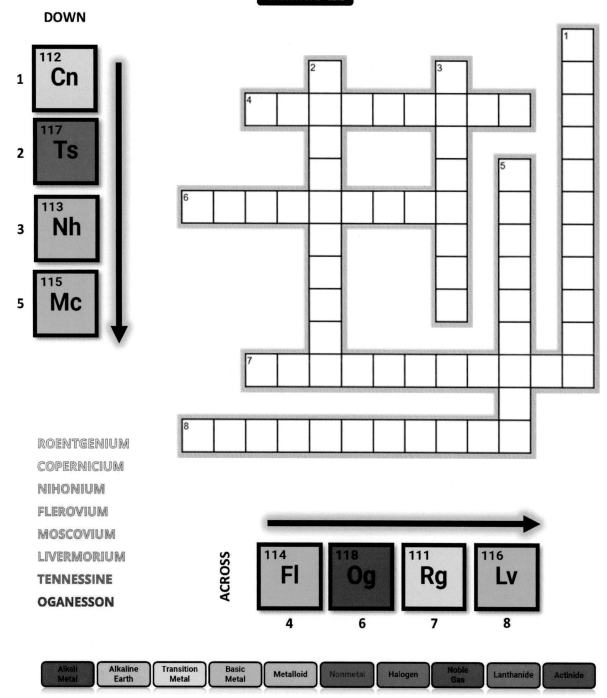

CROSSWORD 12.2

DOWN

1. 112 Cn
2. 117 Ts
3. 113 Nh
5. 115 Mc

ROENTGENIUM
COPERNICIUM
NIHONIUM
FLEROVIUM
MOSCOVIUM
LIVERMORIUM
TENNESSINE
OGANESSON

ACROSS

4. 114 Fl
6. 118 Og
7. 111 Rg
8. 116 Lv

| Alkali Metal | Alkaline Earth | Transition Metal | Basic Metal | Metalloid | Nonmetal | Halogen | Noble Gas | Lanthanide | Actinide |

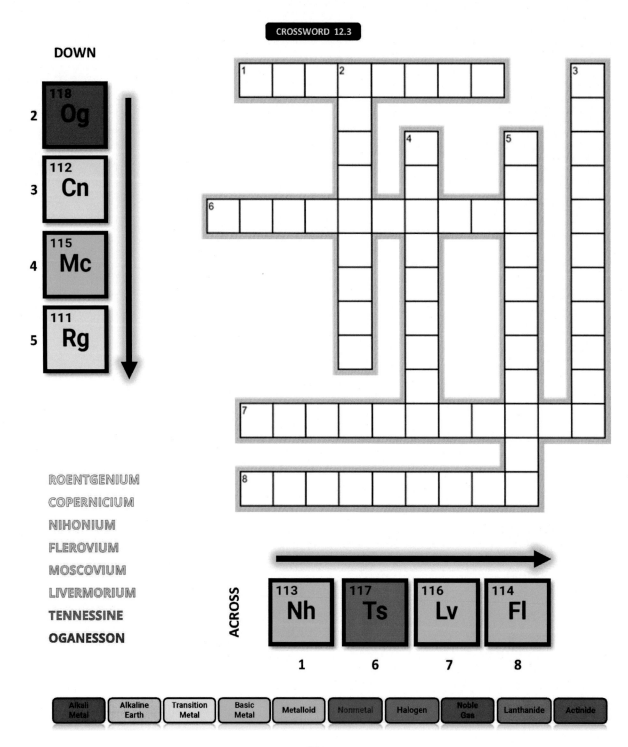

CROSSWORD 12.3

DOWN

#	Symbol	Atomic No.
2	Og	118
3	Cn	112
4	Mc	115
5	Rg	111

ROENTGENIUM
COPERNICIUM
NIHONIUM
FLEROVIUM
MOSCOVIUM
LIVERMORIUM
TENNESSINE
OGANESSON

ACROSS

#	Symbol	Atomic No.
1	Nh	113
6	Ts	117
7	Lv	116
8	Fl	114

| Alkali Metal | Alkaline Earth | Transition Metal | Basic Metal | Metalloid | Nonmetal | Halogen | Noble Gas | Lanthanide | Actinide |

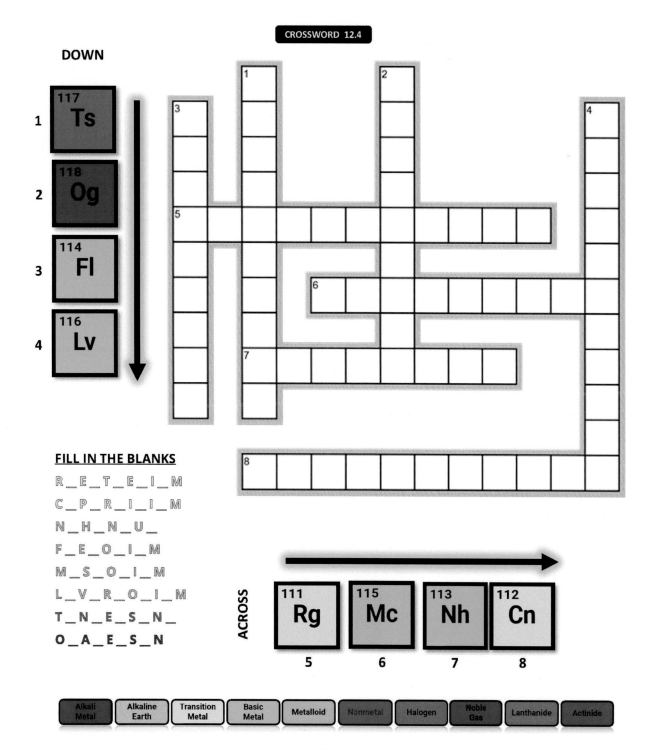

CROSSWORD 12.4

DOWN

#	117 Ts
1	
	118 Og
2	
	114 Fl
3	
	116 Lv
4	

FILL IN THE BLANKS

R _ E _ T _ E _ I _ M

C _ P _ R _ I _ I _ M

N _ H _ N _ U _

F _ E _ O _ I _ M

M _ S _ O _ I _ M

L _ V _ R _ O _ I _ M

T _ N _ E _ S _ N _

O _ A _ E _ S _ N

ACROSS

| 111 Rg | 115 Mc | 113 Nh | 112 Cn |
| 5 | 6 | 7 | 8 |

| Alkali Metal | Alkaline Earth | Transition Metal | Basic Metal | Metalloid | Nonmetal | Halogen | Noble Gas | Lanthanide | Actinide |

59

DOWN

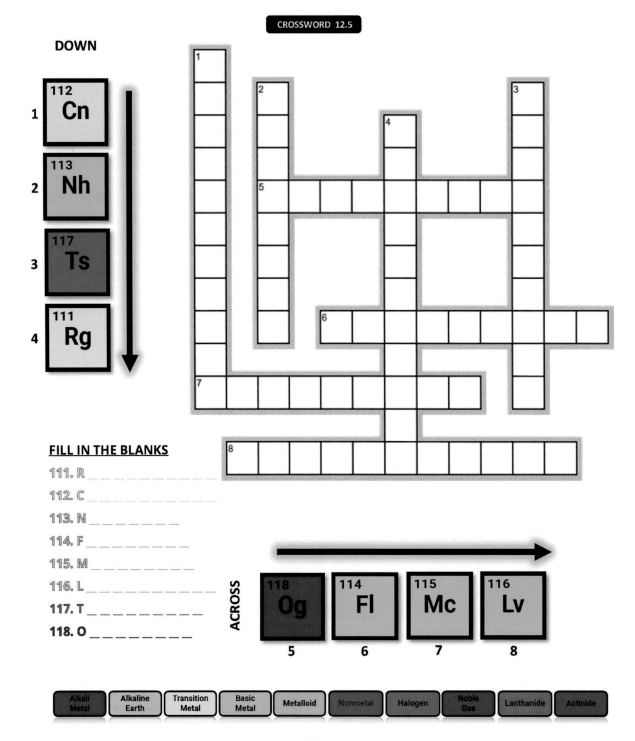

1. 112 Cn

2. 113 Nh

3. 117 Ts

4. 111 Rg

FILL IN THE BLANKS

111. R _ _ _ _ _ _ _ _ _ _

112. C _ _ _ _ _ _ _ _ _ _

113. N _ _ _ _ _ _ _ _

114. F _ _ _ _ _ _ _ _ _

115. M _ _ _ _ _ _ _ _

116. L _ _ _ _ _ _ _ _ _

117. T _ _ _ _ _ _ _ _ _

118. O _ _ _ _ _ _ _ _

ACROSS

118 Og — 5
114 Fl — 6
115 Mc — 7
116 Lv — 8

Alkali Metal | Alkaline Earth | Transition Metal | Basic Metal | Metalloid | Nonmetal | Halogen | Noble Gas | Lanthanide | Actinide

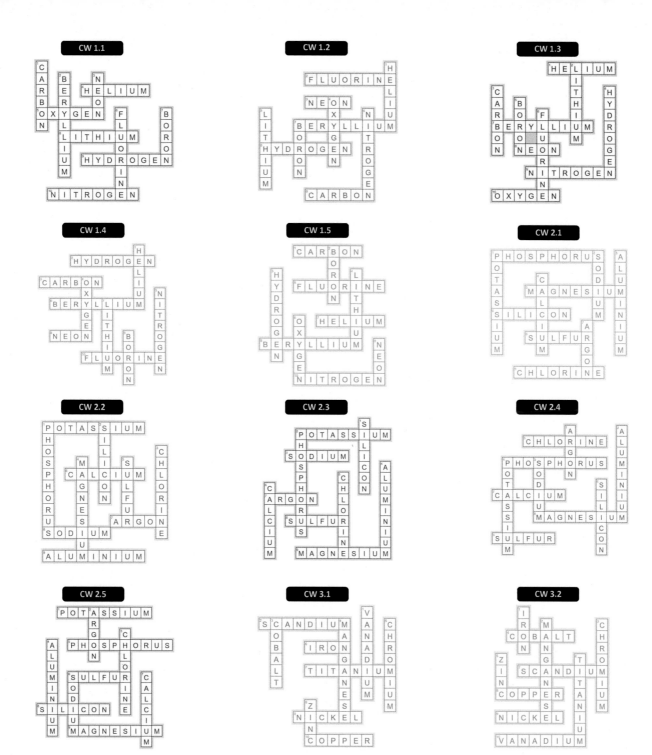

61

CW 3.3

CW 3.4

CW 3.5

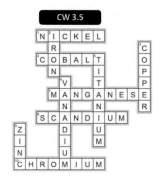

CW 4.1

CW 4.2

CW 4.3

CW 4.4

CW 4.5

CW 5.1

CW 5.2

CW 5.3

CW 5.4

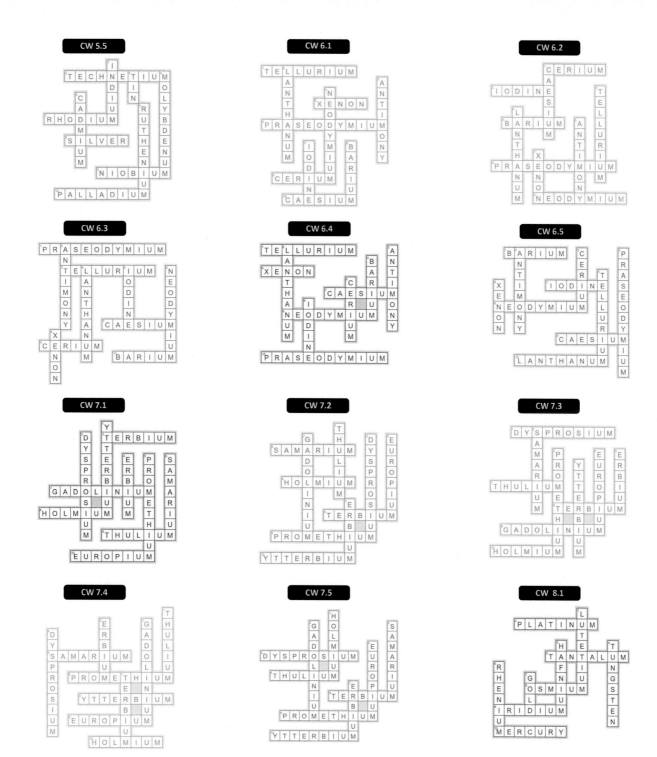

63

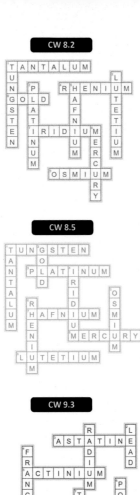
CW 8.2

CW 8.3

CW 8.4

CW 8.5

CW 9.1

CW 9.2

CW 9.3

CW 9.4

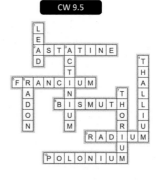
CW 9.5

CW 10.1

CW 10.2

CW 10.3

CW 10.4

CW 10.5

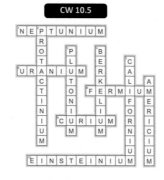

CW 11.1

CW 11.2

CW 11.3

CW 11.4

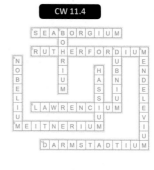

CW 11.5

CW 12.1

CW 12.2

CW 12.3

CW 12.4

CW 12.5

65

1. HYDROGEN	11. SODIUM	21. SCANDIUM	31. GALLIUM
2. HELIUM	12. MAGNESIUM	22. TITANIUM	32. GERMANIUM
3. LITHIUM	13. ALUMINIUM	23. VANADIUM	33. ARSENIC
4. BERYLLIUM	14. SILICON	24. CHROMIUM	34. SELENIUM
5. BORON	15. PHOSPHORUS	25. MANGANESE	35. BROMINE
6. CARBON	16. SULFUR	26. IRON	36. KRYPTON
7. NITROGEN	17. CHLORINE	27. COBALT	37. RUBIDIUM
8. OXYGEN	18. ARGON	28. NICKEL	38. STRONTIUM
9. FLUORINE	19. POTASSIUM	29. COPPER	39. YTTRIUM
10. NEON	20. CALCIUM	30. ZINC	40. ZIRCONIUM
41. NIOBIUM	51. ANTIMONY	61. PROMETHIUM	71. LUTETIUM
42. MOLYBDENUM	52. TELLURIUM	62. SAMARIUM	72. HAFNIUM
43. TECHNETIUM	53. IODINE	63. EUROPIUM	73. TANTALUM
44. RUTHENIUM	54. XENON	64. GADOLINIUM	74. TUNGSTEN
45. RHODIUM	55. CAESIUM	65. TERBIUM	75. RHENIUM
46. PALLADIUM	56. BARIUM	66. DYSPROSIUM	76. OSMIUM
47. SILVER	57. LANTHANUM	67. HOLMIUM	77. IRIDIUM
48. CADMIUM	58. CERIUM	68. ERBIUM	78. PLATINUM
49. INDIUM	59. PRASEODYMIUM	69. THULIUM	79. GOLD
50. TIN	60. NEODYMIUM	70. YTTERBIUM	80. MERCURY
81. THALLIUM	91. PROTACTINIUM	101. MENDELEVIUM	111. ROENTGENIUM
82. LEAD	92. URANIUM	102. NOBELIUM	112. COPERNICIUM
83. BISMUTH	93. NEPTUNIUM	103. LAWRENCIUM	113. NIHONIUM
84. POLONIUM	94. PLUTONIUM	104. RUTHERFORDIUM	114. FLEROVIUM
85. ASTATINE	95. AMERICIUM	105. DUBNIUM	115. MOSCOVIUM
86. RADON	96. CURIUM	106. SEABORGIUM	116. LIVERMORIUM
87. FRANCIUM	97. BERKELIUM	107. BOHRIUM	117. TENNESSINE
88. RADIUM	98. CALIFORNIUM	108. HASSIUM	118. OGANESSON
89. ACTINIUM	99. EINSTEINIUM	109. MEITNERIUM	
90. THORIUM	100. FERMIUM	110. DARMSTADTIUM	

Printed in the United States
by Baker & Taylor Publisher Services